MY LIFE WITH PLANTS

My Life with Plants

A MEMOIR

Sand Mueller

Illbird Press

Copyright © 2020 by Sand Mueller

Publisher: Illbird Press, Tulsa, OK
illbirdpress@gmail.com
Web: www.illbirdpress.com

Design/Production: David Gerard
Cover art: Heath Mueller-Sears

All rights reserved. No part of this book may be reproduced or transmitted in any form or by any means, electronic or mechanical, including photocopy, recording, or any information storage and retrieval system, without written permission from the publisher, except in the case of brief quotations embodied in critical articles and reviews.

First Printing, 2020

ISBN: 978-0-9969621-3-1

Printed in the United States of America

10 9 8 7 6 5 4 3 2

1

The Beginning

Author's Forward

I was born on February 2, 1946. Astrologically, starting with Sun, Moon, Venus, and Mercury, all in Aquarius and with my rising sign in Gemini, a couple of planets in Libra, and I am about as airy as anyone could be. I sort of lived up there where ideas are new. I am not well versed in astrology, but starting very early in my career, I always planted by the moon. I experienced quite a few crop failures when I bucked the heavenly system and planted in fire signs.

My father was stationed in Puerto Rico, which was the center for an anti-U-boat campaign in 1942, but which by 1945 was passed by as the war had progressed. My mother hung out in Miami so she could see Dad whenever he could catch a plane or boat stateside. I was conceived very close to VE Day. This birthdate makes me a certified, bono member of the Baby Boom Generation; in fact, I was very much a big brother to younger boomers who qualified as late as 1964.

This bulge of American children will soon be the subject of many books, and this autobiography will be a tidbit among those many stories. I decided to write about my life, not because I was special or famous, but because my meanderings through the history of the generation took me to some very interesting connections with that history, to many unique locations, and through several different cultures.

I also know that people love plants, and I had lots of stories about them and pretty pictures as well. I hope you find it to be an enjoyable, and perhaps rewarding, read.

September 2020

So the odd part is that I should never have had a career in horticulture. I was a whiz in math, a winner in the 1963 state of Texas high school math competition with a very high SAT score. I was a merit scholar and accepted by the prestigious, math-oriented Rice University. In high school, I excelled in chemistry, and my school team won that state competition also. I should have been an engineer, or doctor, or lawyer, like my friends. But just before I entered college I gave up on math, dropped physics and fled to history. With that decision, I effectively gave up on a successful life in one of the normal pathways to money.

Was it just that I didn't enjoy physics? Or something else? I still do not know.

John F. Kennedy was killed that same year. I reacted to this tragedy with shock and wondered what that meant, what would happen as a result? I remembered Dealey Plaza from when I was a little boy.

My father worked for Sears in downtown Dallas and he, my mother and I went there together more than once. We explored the tiny Neely Cabin and picnicked in Dealey Plaza. This hour or so, there with my parents that warm day, became what I call a primal memory.

Don't many of us hold a handful of memories, like this one from our earliest childhood? They are fogged by time and intellectually limited because our minds are not yet fully formed, but the very early memory retains an evocative vividness that sets it apart from all memories we have thereafter. I have come to understand this after verifying for myself one of the "Laws of Existence" that asserts we have three separate brain centers which do not communicate with one another, but rather depending upon habit or whimsy, take turns being in charge. It is only on rare occasions, when these centers share our attention, that a memory can become this vivid. Four-year-old Sand felt that day in Dallas with his entire self, his being.

I lived with my parents in the Dallas suburb of Oak Cliff. Our address was 2448 Nicholson Drive, a modest new house in a neighborhood full of children my age. Our upwardly mobile parents drove new cars that became more marvelous and powerful each year that passed. My parents and my friends' parents were young veterans from the great war, starting their families in an era of optimism and energy. I did not think about it at the time, but our neighborhood had to be new because the street names were from the Italian Campaign of the just-ended war: Anzio, Salerno, Nicholson, Garapan. Nicholson was a British general and the other three were Italian beaches we assaulted in 1943-44. The true history was that these campaigns had little success and there was great loss of Allied lives, and much suffering also for the enemy and the Italian people. Some of the parents who knew must have been bitter, but we children had no idea what the street names meant.

Probably a couple of years after my Dealey Plaza picnics, I began school. Life changed a lot in that time. School and the world were demanding. The door to the world of innocence and balance was closing.

With my young schoolmates I went to Saturday matinees at the Texas Theater on Illinois Avenue near my home. The Texas Theater was special, and I do remember that place. Once there was a movie about an archer in the jungle. Then, between reels, our hero comes onto the little front stage. He was dressed in khakis and wore the pith helmet of all white jungle explorers. A target was set up and arrows whizzed into the bullseye; then platters were thrown up high above the curtains and our hero struck the flying discs with his missiles.

How could I forget being inside the Texas Theater after that?

On another Saturday, I saw a "coming attraction" that should not have been shown to children which became a profound memory of a darker nature that belongs in a different chapter.

Sand's sister and father

I lived in that house and attended Jefferson Davis Elementary School until the end of sixth grade. Most of us walked to school, and in fifth grade, I became a crossing guard. One rare winter day, a sixth-grade bully pushed my nose in the snow and sullied my crossing guard belt of authority. I was smart and well liked, but felt wimpy and skinny.

In 1958, along with my baby sister, Elizabeth Ann, we moved to Baytown, Texas, where my father became a Sears store manager and an important man in the community.

Winter of 2015. The greenhouses were connected by a passage way. The front was sunken; the back was higher on the ridge, about three feet above the front. I spent $1,000 for the twin-wall polycarbonate in the front. The back was used window glass. The front was 22' x 14', the back 26' x 8'. The passageway added about ten feet by six feet. Altogether there was about 560 square feet of greenhouse. I was heating with firewood and did so for seven years. Propane was better and maybe cheaper.

There are a number of cultivars of purple redbuds. The first one was "forest pansy." This one is burgundy something. I love them.

There were no fans for ventilation. The greenhouse always ventilated beautifully. Because of the westside fencerow, I never even had to shade it. Most greenhouses in Oklahoma are closed in summer because of heat. Look how productive mine was. I had beautiful summer crops from the beds inside.

2

Galveston Bay

Or as it is better called today, the Houston Ship Channel

I was twelve years old around the first of May 1958, when we left Dallas. Baytown, Texas had a population of about 40,000 back then. It flanked the northeastern corner of Galveston Bay on ground just a few feet above sea level. In those days there were many oil wells in and around Baytown and a large refinery as well. There was a clear and occasionally acrid smell of petrochemicals in the air. From my house, I could easily bicycle to the shore of the bay. It was everywhere a shore of mud and reeds; the water itself was brackish and polluted. Local boaters and swimmers went south for cleaner water. The bay here was narrow, and the big ships glided past in rumbling majesty close enough to wave at the sailors. We lived directly across the bay from San Jacinto State Park. There was higher ground on that west side of the bay which led up to Houston's fifty-foot elevation. San Jacinto was the site of one of mankind's most significant battles as the Texans took over the land they had leased from Mexico. This later played a huge part for the entire Southwest of the United States, including California.

Soon I learned that I could bike with a friend to Lynchburg and take the free ferry across the channel to the battlegrounds. There were two nine-car ferries that crossed the 1,000 yards of water. With chuggity motors that sounded like the African Queen, the boats took about ten minutes from dock to dock. Friendly pilots soon allowed us to climb up to their perch above the cars. We got even closer views of the big tankers.

The state park had plenty to recommend it, but the best destination for boys was the Battleship Texas.

Sand and daughter Arena on the USS Texas in the Houston Ship Channel not many years past.

In those days you could go almost anywhere on the ship. It was a huge playground and a museum of both big wars. Being there was like being in 1914 or 1944; take your pick. We bounded from crow's nest to kitchen to the bowels of machinery. We "fired" all the guns. Near the bridge was a display room with huge six-foot models of US Navy warships, the gallant cruiser Houston and the carrier San Jacinto.

I was crazy into ships and naval history in those days. I bought and read the Samuel E. Morrison naval history of World War II, fourteen volumes.

I had these adventures and more, crabbing in the bay, poisonous snakes of every kind, learning to play golf, plus my first girlfriend, kiss and broken heart.

In 1960, we moved across the bay to Pasadena, Texas. My mother, at 39, was pregnant with twins. My father's promotion was to a nearly new, much larger Sears store. Pasadena was leaving Baytown in its dust; it was a rapidly growing community based on turning crude oil into gold. I don't now remember exactly when we moved, but I do remember spending part of eighth grade in Baytown and part at South Houston Junior High School. My new address was 608 Brook Lane, a three-bedroom brick home with an average-sized fenced yard near a creek that they DDT'd for mosquitos. It started to get crowded in that house when the twins were born the summer after my eighth grade.

Aside from making straight A's my main interests back then were shooting hoops and doing Boy Scouts. My comrades in both activities were the Schneider brothers, Carl and Tommy. Carl and I were in the same class, and Tommy was maybe a year and a half younger than I. I was getting tall at fourteen, but Carl reached six feet seven inches in high school, and Tommy was already taller than me. We shot hoops just about every day, and when it was rainy, we glued model ships together with money we filched from our mothers. Carl made Eagle Scout; I never did. He was smart and made good grades, but after ninth grade, he was not part of my circle of friends.

We did travel together with our troop for the 1960 Boy Scout Jubilee in Colorado Springs. It was a fun event marred by theft, bus breakdown, and my case of stomach misery to the point of needing a stretcher. Sometime after we got back, Carl asked if I would let him "cornhole" me. Since I did not know what he meant, he had to explain it to me. I don't recall feeling any revulsion, only a strangeness. It was easy to say no, and I never really noted his reaction. Still, we really were never again friends from that point on. I wonder what the scoutmaster did to Carl?

After ninth grade, dramatic changes occurred. I got a learner's permit at fifteen, and the use of mother's Studebaker. My learner's permit somehow connected me with a remarkable group of friends. We met, initially assembled by our parents, in order to learn bridge, the brainiest of card games. Besides myself, the group included Greg Peters, the fleet of foot. I saw him run the eight-eighty in a high school relay in the rival town of Galena Park. Thousands were cheering, the stadium lights bright, the colorful uniforms flashing, the clear and mathematical red of the track focused into Greg's long strides. He told me recently that he was in a strange kind of zone, one of those hormone states perhaps. His track scholarship to Rice was based on that one race.

Greg and Sand

For two years we were classmates there, but never roommates as Greg lived off campus. Nathan Isgur and Ken Carpenter were best friends. Ken was, like me, a National Merit finalist who also went to Rice, where he excelled and I didn't. Ken died young, in his thirties, I think. Nathan was a Presidential Scholar and went to Caltech. Joel Swanson was a year older than the rest of us. I think Bari Watkins was a year younger.

Bari Watkins was our collective girlfriend, and over four years she dated or was for a time romantically connected with each of us. She was awesomely intelligent, with deep green eyes and bright natural, stylishly cut red hair. Bari was well formed but not a natural beauty; she was attractive and stylish without any artifice. Her PhD was in philosophy from Yale University; later she taught at Northwestern University and founded the Women's Study Program there. She then spent 20 years as a dean and or an academic director at three different colleges. She died of pancreatic cancer at 64 years old.

Rest in Peace Ken, Bari, and Nathan.

Nathan, ah, did everyone else in our group believe you were the brightest?

Debbie Copes, another straight A "member" and very lovely, ended up marrying Nathan. He graduated from Caltech with a degree in physics in 1968 and began a PhD program at Berkeley. Astonishingly for a physics PhD candidate, he was denied a deferment. We probably had the same draft board. Nathan as a freshman had impressed Richard Feynman, his professor. That is Dr. Richard Feynman, Nobel Laureate. He helped Nathan smoothly enroll at the University of Toronto. Dr. Isgur earned his PhD there in 1974. He remained marooned in Canada until Carter proclaimed amnesty twelve years later. His career in physics was very distinguished. Eventually, he came to work in the states and died here when only 54 years old from multiple myeloma.

Joel and I spent a lot of time together even though he went to school 150 miles away at the University of Texas. We both got cars for high school graduation. I actually paid for half of mine with money saved from summer jobs. Joel got a brand new 1965 Pontiac GTO, which was a beautiful vehicle with 335 horsepower, as powerful a statement as one could ask for. I bought an MGB, a two-seater, convertible sports car with wire wheels, red paint, black leather upholstery and 98 horsepower.

We drove each other's car from time to time. With his encouragement, I buried the 120-mph speedometer of his GTO on a classic Texas highway. But my car was much more fun to drive, and I think Joel knew that. Later, my little car got me into sports car rallies and gymkhanas, a thrilling way to beat the shit out of a perfectly good automobile. I often drove like a madman on city streets and highways, and so did most of my friends.

In the summer of 1966, Joel and I took separate chartered student flights to Europe and met in Wolfsburg, Germany, where we took possession of a Volkswagen which sped us through 11,000 miles of the continent in 80 days. We also took trains and hitchhiked since Joel skidded the rear engine vehicle off the road in Czechoslovakia and mashed the fender. Getting out of the country involved an hour's wait in the border guard's office while he telephoned Prague.

We had many other adventures on that trip, but after returning to Texas for his senior year and my junior year, we drifted apart. He graduated from the University of Texas the same year I graduated, since his degree in chemical engineering required five years as did our engineering degrees at Rice.

Joel was valedictorian or summa cum laude, first in his class. From Texas he went to Harvard, where he earned his JD degree and a PhD in business. I visited Joel at his dorm the only time I was ever in Boston, keeping alive our friendship, since neither of us corresponded by letter. Returning to Houston, he started with the Baker-Botts law firm, got married, had a daughter, and bought a mansion in River Oaks; all that before his 26th birthday. I saw him there in 1971.

I never liked Baker-Botts; Baker Oil Tools; James Baker, the secretary of state; or the people they did business with, like Halliburton. I saw Joel for the last time in 1988, when I came down to Houston for my thirtieth high school reunion. We went to a driving range together. He was divorced now; his face seemed sagged to a permanent jowly scowl. He did not seem happy about much of anything or even that glad to see me. I read later that he became a partner in the firm, but never saw anything honoring him as a human being.

During the high school years when our "smart clan was active in blue collar Pasadena," I was also immersed in a most unusual setting in Houston. After some summer tutoring in algebra, I started my sophomore year at St. John's School. St. John's was the most prestigious school in Houston. It was prominently located in River Oaks and loosely connected with the St. John's Episcopal Church. We had weekly chapel, much dishonored by the cynical students. My classmates, aside from the teacher's children and one or two outsiders like myself, were the forty or so richest children in Houston. Outsiders like myself may have been allowed in with hopes we would raise the class test score level as some of the progeny of the wealthy were no brighter than average. None of my St. John's classmates carried the same kind of curiosity and angst shared by my very much poorer Pasadena friends.

Most of my classmates carried the unmistakable aura of privilege. This was more noticeable in the boys. Through two of them I met and played tennis with George W. Bush, who was my age and home from his prep school on the East Coast.

Later when Donald Trump ran for president and his brusque and arrogant manner turned so many people off, I smiled and said to myself, I know Donald; I know exactly in what milieu he grew up and how he came to be the way he is.

I kind of admired the chutzpah and easy success of my classmates. I saw very little of them later, then none, until I drove down from Oklahoma for the thirtieth reunion. I was welcomed to a nice dinner and party at Irv Terrell's lavish River Oaks home. Irv was another of several friends who worked at Baker-Botts from those days.

Most of the attendees were successful and wealthy. Some seemed sad, disappointed in life. I was treated respectfully and they particularly liked my descriptions of living in poverty and trying to be self-sufficient.

Wait. I'm getting ahead of myself.

From when I turned sixteen, I worked summers at Sears and earned $1.25 to $1.50 an hour, plenty of money for a young lad to fix up his Studebaker. But for the summer of 1965, my father arranged with one of his Rotary Club buddies for me to work at the Diamond Alkali chemical plant on the ship channel in Pasadena. I think about five of us were hired based on this favoritism. We earned a union salary though we did not pay dues. Somehow, I did get a union card, which I proudly kept around for years. That salary amounted to $3.41 an hour, more than twice what I had earned at Sears.

There were tens of thousands of Black and Hispanic men for whom that pay would seem like a fortune, whites as well. My first day on the job, I helped clean up from an explosion that had killed two union members.

Then day two, I was sent to the maleic acid factory. This was a relatively small, about a football field-sized, part of the refinery that made feedstock chemicals for the manufacture of the earliest PVC pipe and agricultural plastic. I actually never saw anything plastic in my summer at that place! Our part had a central reaction chamber sort of like a big globe 25 feet across. Feeding into it was an 8-foot or 10-foot or maybe even bigger pipe through which was pumped natural gas or other combustible. Smaller pipes would have injected into the chamber the various catalysts and chemicals to make this shit.

Truth is there were pipes all over the place going God knows where, and they all had valves which the head operator needed to know about, not me. There was also a very large cooling tower with three giant fans blowing steam seemingly out of the ground. The vapor they blew up did not seem toxic; it wasn't exhaust, just plain heat. There was a wooden structure around these fans with a deck on top. I used to go up there and sunbathe on my lunch break. You couldn't hear anything because of the noise, and the whole edifice did a vibration massage to my reclined body. I fell asleep up there once and was late getting back to my duties.

The final main component of our factory was a 2200-horsepower, diesel air compressor, the same size and rumble as a locomotive engine. It pumped air into our reactor chamber to combust everything inside. Maybe that reactor was 50 feet across, not 25 feet, jeez.

There was a lot of horseplay at work, led by the 40-year-old manager of my shift, Johnny Johnson. He was what everyone then called a "good old boy," not educated, but he had his union card and was plenty competent enough for the job. He wanted me to date his daughter; she was way too fat. We worked rotating shifts, that is to say, day, evening and graveyard. So the attack time was either early AM, mid-afternoon, or at midnight. The weapon, a fire hose.

No one could walk from clock in to our working office without passing through a hundred yards of peril. Traps with buckets of water held by an inch of ledge were also set. Once our erstwhile victims swarmed in silently and early, seeking revenge as we exited our day. I drove home wet, haha.

On a morning at the very end of my summer, I came in at 7 a.m. and sirens were blaring. Johnny was already in the office monitoring his many analogue and mechanical gauges. I found out that the big diesel air pump had been going out all night. Now we were shutting down the plant. All the lines into the reactor had automatic valves to shut off feedstocks. You could not tell if they worked or not, so there was a hand valve, a wheel three feet across, of heavy steel with tight threads that took many turns to seal and cut the flow of combustibles. This was mounted directly on the side of the monstrous, now simmering reaction chamber.

I took about five minutes to close this valve, sweating in the heat of the place and considering that this unstable cauldron could blow up at any moment, repeating the death scene from my first day. This five minutes was just about the only serious work I did at the plant that whole summer besides read gauges and inject oil into little electric motors.

At home I am sure I also played some tennis with Greg and Joel, and golf with my mother and father. That and some desultory dating and zipping around in my new MGB completed my summer.

In 1966 I went to Europe, so that the summer of 1967 was the last that I worked on the ship channel. My new job was at the Rhome and Haas chemical factory, also on the channel. In fact, aside from a wharf or two for dry goods or machinery the entire ship channel between Lynchburg and Houston was lined with these plants that used oceangoing tankers for transport of their ingredients and concoctions. My pay at Rhome and Haas was a more modest $2.50 an hour, still a good wage. But my job required much more. I was a painter's helper.

I wish I knew what R&H was making besides chlorine gas, the mere whiff of which caused concern. One choking afternoon every worker at the plant had to flee the gas that had suddenly enveloped the plant. Everyone just ran away, helter skelter; we didn't know where to go. Nobody died from that thankfully.

My job at the plant was to wire brush and clean surfaces for the painters who followed. Outside it was actually kind of fun straddling the two-foot diameter pipes that went from the tops of storage tanks in their network of flow. We scraped and scooched backwards about twelve feet above the ground. This work made one strong and agile, but I don't think breathing the dust did much for my health. In fact, our crew had monthly urine checks to monitor our toxicity level.

The first task we had that summer was not outside in the tank field, but in a giant warehouse that housed what looked like the cut-flower beds I worked in later. These were four feet wide by about 100 feet long, metal lined troughs that stood about two feet high with at a 30-foot walkway between each bed. There may have been 50 or 100 of these troughs and they were filled with shimmering mercury. I don't think the mercury was that deep, but supposedly we had one-half of the world's entire stock of mercury in these troughs! This mercury was constantly being bathed in 15,000 amps of electricity. Working with a couple of other employees, my job was to mount a ladder up the side wall to some catwalks and on to a scaffold that was hung from the ceiling at a height that would allow us to hand scrape and wire brush the metal-surfaced roof.

We wore long pants and shirts, gloves, hard hat and a respirator. It was about 25 feet above the beautiful shimmering mercury. It was roasting and back-breaking work. We collected our dust and debris on a tarp under our feet. Since part of our job was moving the scaffolding, we carried wrenches in our back pockets. Each time we came down or went up, we walked down one or another of the aisles, and as we did so the wrenches in our back pockets would pull or jerk from side to side drawn by the magnetism of the surging electricity. Later, I saw from a distance, before the ambulance came, an electrician who had taken a shock. He was blue. I heard that he lived.

I lived that summer in a poorly air-conditioned apartment with my first love so there was a lot of sex and sweat, living and working near this once beautiful bay.

Asian lilies! The orange is a little reseeding coreopsis. The coneflowers were originally Pink Parasol, but seedlings are coming with a lot of white and downturned petals (parasol). So undoubtedly, Pink Parasol is an F1 hybrid, and we see the parents coming back to say hello.

The previous summer photo showed a dead cherry branch. That was caused by an 80-mph derecho. Never had them when I was a kid; now they are worse than tornadoes. If the wind had been out of the east instead of west shown here, my greenhouse would have landed in the next county! Luckily few big winds come from the east here.

This was my excavation. The raised beds you see to the left. I laid pallets over them for potted and bedding crops. When it got in late spring, my container plants went outside and I grew wonderful summer peppers, tomatoes, and melons in them. In winter I had success with greens, herbs and even peas.

3

Seven Years of School

Five years after my sixth-grade move to Baytown, I was a senior at St. John's School in Houston. St. John's was, and is, a K-12 private school for the wealthiest and most privileged children in Texas. I was there via a chance meeting of my high test scores, my mother's wish, and St. John's desire to admit one or two smart outsiders ready to pay their tuition, $700, and to help raise the aggregate test scores. It was the fall of my third year there, my senior year, when bustling with my classmates to the cafeteria for lunch, I heard the excited rumor of Kennedy's assassination. There was very little time between "The President was shot"... "The President was rushed to the hospital"... and "The President is dead."

Among my generation some form of this memory is almost universal.

Later that day Oswald was captured in my Dallas boyhood theater, the Texas Theater. He was shouting, "I am just a patsy. I am just a patsy." The shock of the President's assassination left yet another memory. It is a stark memory not just because of what it was, but in contrast to all my other memories which had become dull and faded as I passed through childhood and puberty. It flowed through my childhood memories of Dealey Plaza. The murder expanded with the news coverage, the photographs, and the Zapruder film; the articles and books even back then multiplied their effect on me. In Texas, mistakes by the cabal had slipped through and become narrowly public. I was pulled by a fascination which continued and broadened through the other murders of Malcolm, Martin, Robert and so many others, including those of persons who knew too much about JFK.

Three years later I spent a college summer in Europe. In Houston, believers in "conspiracy" mostly whispered to each other, but in Europe there were loud dissenting voices from all the European intellectuals and students. Upon my return I began my junior year and became an angry opponent of the Vietnam War and a college politician. For the next two years from within the safety of the college milieu, I publicly accused our government of criminality. When I led a demonstration against CIA recruiters on campus, my protest reached the Houston Post, which accurately described me as "red-bearded Sandy Mueller."

My notoriety and the atmosphere even at conservative Rice propelled me my senior year to the presidency of my college, and my radical reforms were enacted by the Baker College cabinet. How ironic that I was president of this college of 300 men built by and named after Captain James Addison Baker whose family power was fueled by Nineteenth-, now Twentieth-century oil.

The Baker College Hall was built in 1920. It feeds 300 with kitchen and monster foyer; it is one of the most ornate, collegiate, and beautiful buildings imaginable. My name in bronze still hangs in that foyer today. The Baker family from those days has been firmly allied with Rockefeller, Rothschild, war, and pollution.

My "reforms" based on campaign pledges were, one, we could now keep and imbibe liquor in our dorm rooms; two, women after they signed in were allowed in our dorm rooms; and three, our conservative and unpopular college master would be replaced with someone more open minded. The university provided a house for the master and his family. Dr. and Mrs. Wischmeyer had two children whom I had known at St. John's. They were one and two years behind me. The university had written in their by-laws that the masters' positions were temporary and not intended to last ten years. Dr. Wischmeyer was the archetypal hindrance to our modern way of thinking, a most conservative man and too shy to relate to easily. I spearheaded all my initiatives and the university and the Baker College Cabinet went along with all of them. Perhaps the university's decision based on my (our) callow testimony was the correct one.

I think masters are rotated now, but the Wischmeyers did have to move out and lost their free meals. Writing this now I cannot recall wishing ill for the Wischmeyers, my supposed friends, to get kicked out, but that was a consequence of my actions. I guess I sought what I wanted and didn't consider the consequences.

Wasn't I special and powerful? I squirm under the gaze of this analysis even now.

I had lots of friends among faculty and students. I also had my detractors. That surprised me. I only won my radical presidency by a vote of 115 to 114, and my votes for outstanding senior were the lowest of the ten selected. Didn't everyone want an end to war and brotherhood among men? I've flailed at a lot of hornet's nests since those days and have been painfully stung over and over. Now in my fading old age, I have plantain salve.

In 1967, Muhammad Ali came to the Rice campus. He wasn't invited to speak in an assembly hall; he had just been indicted for failure to report for duty and stripped of his crown. No, he slipped on campus; people were gathering in our Baker College kitchen where he was talking to mostly black kitchen employees. I rushed into the crowded room. I remember some of what he said was about his Muslim faith and included the bluster he was known for.

Then a young woman in the audience spoke up as he took questions at the end.

"Muhammad, I have known some fighters and their faces were cut, scarred, or swollen. But your face is unmarked, how do you do that?" The Champ completely changed his demeanor and with direct eye contact with the young woman, as if no one else was in the room, talked about his training, the people who helped him along the way, but above all that the creator had endowed him with such a quickness that he rarely got hit in the face. He may have been flirting with her, I don't know, but I had always liked Cassius Clay from when I first saw film clips of him winning the Olympics. Now I liked him a whole lot better as Muhammad Ali, the brave Champ.

My lover and I went to a local nightclub early one evening. We were drinking beer, almost the only patrons when the musician we hoped to enjoy came into the room.

I called out, "Hey, Lightnin', can I buy you a beer?"

He declined the beer and came and sat at our table. He had a strange appearance to me but kindly, very kindly. We talked for a good while. I only remember the last topic, old? How old? It held the aspect of life experience in it. He was only 56, but he seemed old, wise. Lightnin' Hopkins was the first Black man I ever talked to aside from the "colored" waiter at our country club. I know he was considered by some to be the finest blues musician ever to strum a guitar. Lightnin's music was just as strange to me who had grown up listening to Artie Shaw and Lawrence Welk. I still listen to Lightnin' Hopkins from time to time. There was a lot of different music coming on the scene in those days, our world was permeated with it and we did our homework with the turntables circling.

In high school, I had been all-conference goalkeeper on our championship soccer team. I continued playing in college, but not at Rice where there was no NCAA team, but rather at Memorial Park in a club league. There were few local boys like me playing. Everybody was from some other country. I played with Englishmen, Manchester United! Of course there were Mexican teams, but also Arabic, African and South American. Games were played every Sunday except in summer. I was considered a damn good goalie, and our team won the title two of the three years I played.

Just about my last game, as champions, we had an exhibition against Houston's first professional team, the Stars. At the half we trailed only 3 to 1. They were, of course, bigger, faster, stronger, and younger than we were. I was a pretty good punter and equal to the other goalies I had faced. The Stars goalie was a Dane, six feet four inches tall. His punts came deep down into our half of the field. On one of them came a breakaway; their big center forward had gotten the kick behind the defenders and had an unimpeded opening to the goal. He was dribbling as fast as possible toward me and the goal. The goalie in this situation has only one play, to charge directly at the attacker to narrow his opening; "He's going to have to kick it before you collide." The previous year I had gotten a bloody nose on this exact play. Bam, came the shot. My rigid arm caught the ball; no wait, the shot was so hard it blew right on through into the net. Final score: Stars 8 United 3. I loved playing soccer, and like many other college "stars," my career ended with graduation and moving on.

As I mentioned before, my college grades were indifferent. Actually, my efforts declined a lot after two freshman classes. The first was European History, my major subject. There were about 400 in the class. Grades came from two tests and a final. My two tests were 91 and 84 which put me near the top of the class and I had an A- after these two scores. I studied like crazy the whole semester and for the final. I got a B+, which was my semester grade. OK, fair enough, but wait, there were only seven A's out of 400. I was probably in the top ten; I had worked way too hard in that class to end up with a B.

A's now seemed out of reach. In freshman biology, I quickly saw that I was taking essentially the same course I had aced in high school. Nothing in the course was new to me. However, the ancient professor connected this knowledge with "key" words that he made up. It turned out that if you did not use the key word from his lectures your answer was wrong. I got a D in that class. After that I did just enough to get by, only making an honor roll my final semester when I squeaked in with all B's. I may have graduated in the bottom quarter; I never knew my exact ranking. It was brought to my attention later that there were a lot of Rice student suicides back then. I don't know that to be true, but it is food for thought.

My brief peak of fame and power came to a sudden end upon graduation. My history major held little interest to industries or government agencies that got deferments for their employees. My soon to arrive new draft status was 1-A. I had thought about my alternatives. Selective Service had a policy they called "channeling." By dangling different deferments, young men could be channeled into pursuits approved by the government. Peace Corps was the most famous of these channels. I made the phone calls and did the paperwork to become a Teacher Corps intern, and on June 1, 1968, the day after my graduation I took a flight to Washington, D.C. and from there to New York City. This was to confirm my application to be a teacher in the Big Apple. There was a reason I had chosen New York. I had gotten married in April, and my bride was employed by IBM in Poughkeepsie, a train ride from the city. We had met on our flight home from Europe. She was a year older than I, and on a warm fall night, on a blanket, set on a lawn where few people went at night, she ended my virginity. Bless her heart.

This was the glasshouse as it neared completion. The beautiful casement windows on the right were never suitable for a roof, but they were what I had. I did repaint and caulk them after three years. They lasted five. The heavy sliding door glass panels on the left are the perfect roof, but plenty heavy. I have 18 now if there is to be another greenhouse.

It's like a magic fairy tale inside at this moment. I only heated the greenhouse to keep it above freezing inside. Almost every plant will thrive with nights this cool. Citrus and succulents loved it. Importantly, I ventilated on sunny winter days and kept day temps below 70 degrees. This is the fatal flaw of hoop houses, indeed most greenhouses. Insect pests love those 90-degree winter and spring days. Nature's guidance is followed so that the plants inside live in harmony with outside, just warmer is all.

Red bee balm or Monarda. The lilac-colored flower behind seemed like a kind of statice, except that it's not.

4

Sexuality

It always seemed to me that I was the last to learn about sex. I was eleven years old when Tommy Bartlett my older cousin told me how babies are conceived. I was shocked and disturbed. I did ask my father about it later, and I am sure his response was thoughtful and considerate. He may have mentioned the church. Mother, Dad, and I were devout Episcopalians. I was an acolyte and loved the ritualistic services with music and incense.

I wonder how many of us felt a connection between religion and the sexual act? Since my church ordained that we should remain virgins until our wedding night, my mind now connected the two with judgment. I wonder how many hurried marriages came out of this common religious doctrine?

My father and I never spoke of sex again, and my mother certainly never mentioned it. Over the next nine years all my education on this one subject, which increasingly interested me more than any other, came from my teenage friends, Playboy magazine, and the movies. I faced this most important of questions with an undeveloped psyche and some subliminal inputs that lurked behind every girl I wanted to kiss.

One of those inputs had been formed in me back at the Texas Theater when I was six years old. We were watching Buster Brown or some other children's classic when coming attractions played a clip from a trailer for the movie, "The Nine Wives of Bluebeard." In the clip, Mr. Bluebeard was drowning one of his wives in the bathtub. It showed her little feet kicking as her life was submerged. I take this scene to be the source of a foot fetish I retain to this day.

Curious, no, how something imprinted at random could compel my attention toward a woman's bare foot and do so for the rest of my life. It opens a question. How is personality formed? What is the relationship between desire, judgment, and restraint? Women commonly paint their toenails and adorn their feet. Surely this is a signal for attention, and "attention" is where I am. On the other hand, women, quite correctly, may take offense if someone is "caught" staring at their feet. Who draws the line between desire and restraint? Fortunately, my fetish never put me in front of a magistrate. Overall, my bodily senses enjoyed my fetish and so did some of my partners.

As for the portrayals of lovely young women being murdered. Really who wants to see that? What six-year old should see that? Films are not real life, but does a six-year-old understand that?

It's embedded in my personality now. What must I think about that? Actually, when viewing a movie or TV show with murders of women, I experience a split between titillation and revulsion. This may verify a quote made in the 1940's from a Hollywood magnate who said, "If we show them the right images between the ages of one and six, we can split their personality in two." It seems like that happened to me. It doesn't feel right or in harmony.

Reality is attending the death of someone with being, someone you love, my father. This is not to be missed. It is a unique sacred event which reminds you of yourself and your inevitable death. It also honors your ancestry and that higher place with so many names. Reality is also the sacred event of copulation wherein love and gratitude for your partner multiply the physical joy felt in your body. This reality honors the Earth and the human species. It is the crowning achievement of our lower nature.

I had a poor body image. I sincerely believed I was too skinny and too shy to be attractive to the opposite sex. I addressed this doubt, not by withdrawing but with determination to overcome it. After a single hard-won kiss in sixth grade, I began dating when I got my first car in 1962. This vehicle was a 1953 Studebaker Starliner Coupe. How I wish I had it back today! Everyone knew that this car was special, but in my teenage milieu, it was underpowered, slow, like me. I flogged it unmercifully. Over the next four years I had a couple of touches but no scores. One thing was obvious; while I might seem to be in line with the church's doctrine, it wasn't because of it. I knew that if the opportunity for sex came, I would follow my desires rather than any scriptures. My shyness and ignorance kept me in slow motion, so the agony of virginity was going to last a while. Looking back now I suspect that my falling away from the church may have been set in motion by my response to this contradiction between virginity and passion. Of course, I was such a perfect child, spoiled, Mama's Little Prince, I wanted to do what I liked and not be judged by anyone.

So in the summer of 1966 with money saved from the previous summer's lucrative job, I flew to Europe on a student charter and spent eighty days there traveling with Joel who had the same virginity problem. We had chosen a few fixed stops, but our schedule was mostly open to what we hoped was new and beautiful. Trouble is we mainly went to museums and stately churches. Our pre-trip destinations were at relatives of his in Sweden; Le Mans, France for the 24-hour car race; the Grand Casino in Monte Carlo; the beaches of the French Riviera; and the nearby island of Hyeres. This last, lesser known destination had been spied reading Playboy magazine in our dorm room; it was the site of the world's largest nudist colony.

I'm not sure how we thought we were going to get laid at the Casino where our cheap suits and youthful inexperience surely stood out among the elite socializers gambling in their tuxedos and designer dresses. Nor would we fare better the next day on the beaches from which we stared at the beautiful people swimming from their yachts just offshore. The ferry to the nudist island also went to an island nature preserve and carried forty-ish German men in flip-flops along with matronly couples in hiking gear with binoculars.

Once on our island, we docked and were given a basket for our clothes. Only the first-timers like my friend and I kept on our underwear on as we left the entry. The path led onto an outlook on a cliff about fifty feet above the beach. There were about a score of nudists on our level, but most of the hundreds were on the beach below. In their midst was an old man whose testicles hung deeply. He held a large mirror in his hands, and using it, he reflected sunlight upward onto my friend and I. As he did so, many on the beach and around us began clapping. It was the first and only time I was ever applauded for taking off my pants!

We spoke with no one while we were on the island and even the sight of the most beautiful naked young women failed to stir our obsession; we wanted no further applause. The day's experience did not open the pathway to an orgy we would both gladly have attended. That fall after my friend and I returned home for the next college year, we finally both scored, I at the hands of a smart, lovely and wonderful girl who I married a year and a half later in the spring of my senior year. Unfortunately, this experience and relationship did not erase from my psyche the fire of testosterone untamed and disrespected.

Spurred on by the same media that split my personality and gave me my foot fetish, my selfish desires led to infidelity in marriage and masturbation. Some literature describes my good fortune in not satisfying this impulse until I reached twenty years of age. I am no poster boy because of delayed gratification; it didn't seem to help.

Othonna crassifolia, "Little Pickles," had endearing flowers. I grew this plant steadily for about seven years, then never saw it for thirty years, then ordered some with the internet. It was very successful for me, easily one of my $100 plants (ones that earned me that much from sales).

Viola, "Johnny Jump-up," is an old fashioned perennial flower often planted along walkways or even between stones. I only grew it for baskets. It was always hard for me to find bed space outside in the sea of Oklahoma grass.

5

Teacher Corps

My first news from the Teacher Corps was bad and came quickly. NYC schools only took interns from those returning from Peace Corps. They suggested that Chicago was taking interns, and thankfully my hurried application was accepted there. I joined 75 other recent college grads in, what to me, was the scariest city on Earth. I had visited my cousins there when I was nine years old. My bride and I had one month together in Poughkeepsie before I started in Chicago. That month was happy and eventful. The scenery was fantastic, how lovely, the Hudson Valley. We drove over the bridges through towns each more iconic than the next. We visited the mansions of Roosevelt and Vanderbilt and up in the gorgeous Catskills; we took in artsy, crafty Woodstock, a year before the famous event. Later on our trip moving me to Chicago, we saw mind-boggling Niagara Falls and the beautiful, clean city of Hamilton.

During the day when my wife was working, I volunteered for Eugene McCarthy's campaign to win New York state's delegation to the 1968 Democratic National Convention. I remember enjoying that, but hardly anything of what I actually did, walking around some, making a few phone calls, sticks in my memory, except that there was pretty scenery everywhere. McCarthy did win the New York primary, but when Kennedy trounced "Clean Gene" in California, the nomination was in his hands. All of us McCarthy people were ready to throw in with Bobby, for about an hour anyway, then television was saying, "… walking backstage to exit the hall," "Secret Service with him," "killed by mad Arabic gunman."

Our fourth hero killed, every murder an obvious cover-up.

After my wife drove me to Chicago in July and till September, I lived in the men's dormitory at Concordia Teachers College in the suburb of Berwyn with my new intern friends. When we went into the city, we drove or took the "L" train through the West Side where close to one million Black Americans abided in poverty and urban decay. From the expressway, we could see burned out buildings left over from the riots after Martin Luther King's murder four months previous. When school started, the dorm closed, and we found our own housing and were assigned to our schools.

I moved to an apartment building near the "cool" and "new age" integrated neighborhood of Old Town in the Near North Side. It turned out I was the one who integrated the apartment building I first lived in. My school was in the heart of that same West Side we had seen from the "L" train. These, what we called slums back then, had a noticeable smell of poverty. Langston Hughes Elementary had 1,100 students, two of whom were white. Each school in the program consisted of a team of four or five interns and a team leader. All the team leaders were experienced black school teachers. We were like student teachers, all with college degrees, but none in Education. We would finish in two years with master's degrees in Urban Education.

In retrospect it was a good program. We worked with several teachers, all of them dedicated and skilled. We got to do a good bit of monitored teaching. I had a garden growing in cardboard boxes in the windowsill and a 300-foot-long display of the planets to scale in the hallway.

Discipline could be a problem; it was for me. There was social outreach and once I took my oldest sixth-grade boy to a Bull's game. There were lectures and work in the community as well. Late afternoons we attended classes for the degree. We were immersed in the Black culture, and we found it to be good. We found it to be rich, long-suffering, and fascinating. Unfortunately for me and the other Caucasian interns, some parts of the culture did not like us or want our "help."

At home my apartment neighbors were all Black, and before long I was accosted and humiliated. In broad daylight in the entry, he stopped me and demanded to see my identification. I meekly offered my wallet and he knocked it to the ground more or less cursing me at the same time. He didn't hurt me, but one of the speakers we heard later said that if we wanted to help, we should kill ourselves. This sure was a different world from Houston and that was only four months ago.

The 1968 Democratic Convention

Everyone I knew was against the war. Three months previously I had been a student in Houston. Now suddenly, I was in Chicago for the high-water mark of the antiwar movement, the 1968 Democratic Convention. It came shortly before we moved out of our dormitory in late summer. Bobby was dead, my man McCarthy had the most delegates, but not enough to win outright.

I don't remember when we realized that Hubert Humphrey would be the nominee. McCarthy was an outsider; and the insiders didn't like him. (Things don't change much.) Young activists against the war had gathered in great numbers from around the nation, and they were sleeping in the Chicago parks.

Mayor Daley's police attacked the night before the convention began. Who knows where those thousands fled to for the rest of the night. Many were arrested; many were clubbed. I suppose we saw it on television, but if not, we read about it in the paper on our way in to town the next day to participate in the big event. The convention was starting that evening. Tens of thousands of us gathered at the band shell in Columbus Park just across the Illinois Central tracks from the convention. There was music and anti-war speakers exhorting for something. There was no leadership; most everyone was there for the event. Event it was, a sort of a military parade. There was a command post on top of the Field Museum complete with antennas and a machine gun, tanks with cannons aimed at us on every bridge over the tracks, and a phalanx of soldiers and police shoulder to shoulder from the Congress Hotel to Michigan Avenue. When they were fully deployed, not too long after we arrived, they began lobbing tear gas into our crowd. We were forced north all the way to the Chicago River. By the time we got there just ahead of the tear gas, little ambition remained for heading back south towards the convention. Only a head cracking and jail visit was offered in that direction.

Humphrey's nomination meant that the war would go on. The national effort to stop it dissolved into each individual's struggle to stay out of the armed forces.

A flat of tomatoes grown in my unique paper pots, "the best bedding plant container ever."

The blue one is an agave, the variegated an aloe. The tall Cleistocereus is called Silver Torch. The baskets are gazania and calibrochoa. Fennel is in the net and a blonde bell pepper is peeking behind the Silver Torch.

Early winter vegetable beds with gorgeous petunia basket and a bit of black-eyed Susan in the corner.

6

My Friends

Fortunately, staying out of the armed forces wasn't a problem for any of my friends who were teaching with deferments. I was the only one out of a close circle of about ten with status problems, and they were all cheering me on to avoid the Selective Service's clutches. I had a bad lottery number, 144. My three closest amigos were Mike Polak, Vince Pascale, and John Laue. There were two Lenny's and with them, rotating girlfriends.

One Lennie came and spent time with me at my parent's home on Lake Livingston, Texas, where in the summer of 1971, we water skied for hours and hours. I lost track of the other Lenny when he moved to Northern California in the early 70's. Apparently, he's been growing pot up there ever since. In 1972 when I remarried, Vince, Mike and John were the "men" of my wedding and handsome in their tuxes. Vince occasionally procured LSD and/or synthetic mescaline. I never knew the difference.

One memorable night, with Marsha sober and driving, we picked up my three brothers from John's South Shore apartment. I cannot remember when we took the acid. Driving from there, we headed to a Sha Na Na concert in the big events building south of downtown. We were cruising in the huge 1965 Cadillac my parents had recently given me, and then, with the temperature outside a frigid 16 below zero, my friends, just the three of them, Marsha did not join in, with great laughter began racing their power windows. There were several rounds and twos out of threes. I think the back, passenger side won, congratulations to Michael Polak. Anyway, that became one of several legends of our friendship.

We all liked being in nature and often met in the Indiana dunes, also camping in Turkey Run State Park in Indiana and at the La Crosse, Wisconsin, Octoberfest where I confronted a motorcycle gang ruining everyone's sleep. They in turn ran their bikes offroad through our little encampment, knocking down our stuff and almost pulling our tents over. And lastly playing one of our very competitive games of basketball under the six-lane highway near my house, I elbowed John in the mouth as I drove in for another missed layup. John had crooked teeth, and the one I got used to stick out. He later always thanked me and said it made his mouth much better.

Mike married Paula, they had two successful children, and are both retired after long careers in the Chicago Public Schools. Now they live in the wooded nature of the eastern shore of Lake Michigan. Vince had a large vacation home on the Indiana dunes where we often gathered. His father was a wealthy and domineering heart surgeon, and Vince was the oldest of ten children. He actually grew up in the wealthy suburb of Olympia Fields in far South Chicago. Sadly, and curiously, I knew only two men my age from there and they both died young from rare aggressive cancers. Vince got a sore back that was eventually diagnosed as a cancer that seemed to go up his spine into his brain and back again. He died on the operating table in 1983 at age 37. John grew up in the dunes in Miller, Indiana. His father was an editor for World Book Encyclopedia. My 1954 set was a huge part of my student days. In fact, my daughter Arena has my entire, now 66-year-old, set in one of her bookshelves.

The Laue's fairy bread house in the deep pine shade of a dune had a yard covered with green velvety moss. John finished his master's degree and moved to Los Angeles working in the schools, first as a teacher then as a drug-and-troubled student counselor.

He married and had one child, but soon divorced. He is kind of an enigma because despite being a life-long student of Buddhism, associated with Ashrams even, he also had serious bouts of addiction. He remains a firm practitioner of the 12 Step program, but oddly enough, the one main group he attends is the 12 Step for sexual addiction.

I also attended 12 Step programs twice for pot and also for sexual addiction. I was kicked out of my first group by dry drunks, me being the only non-alcoholic guy. I believe I did benefit from going, but my whatevers were never as debilitating as those of some of the others. John is a thoughtful meditative man of great virtue. Mike, John, and I rarely saw each other over the years, but we have stayed in touch and remain close. Only Greg now is an older friend.

I often found gazania baskets in the box store reduction bin for $3 or less. Gazanias go in and out of bloom in a remarkable way. The store assumes one flush, and then sells them at a loss. When they return to bloom, wow. I also grew them from seed. They overwintered well in the cold greenhouse, but I never kept them for long; they sold.

The begonia grew well for me, and I propagated it a little. It lived in the house; no 33 degrees for it.

A good year for this five-gallon citrus limequat. The skin is sweet and edible, but the fruit sour. Citrus were all over the place for me, good years and bad news. One year, a horrible attack by mealy bugs, the next, clean as a whistle. I believe the longer you have them the bigger and stronger they become. I took care of a 55-year-old calamondin orange in ten-gallon was five feet tall and six feet across. One year it yielded 11 lbs of the tiny, sour citrus. Made marmalade. Christmas gifts for 16 people.

7

Marriage, SSS, Black Chicago

My wife had applied for a transfer from IBM headquarters in beautiful New York to suburban offices near Chicago. It was an unhappy move for her, and in our new life nothing went well. We went on long bike rides along the beautiful Chicago Lakeshore shore. Too far to the south we rode, and then in an otherwise lonely spot, we were suddenly surrounded by about a dozen "sixth graders." We desperately escaped though I got a weal on my back from the chain they stole and whacked me with. It seems like life was not going to allow us to opt out of the history of white people brutalizing or murdering other ethnic groups. Humbling certainly, "I am a good person, certifiably not racist," will mean nothing to anyone who doesn't know you personally, and they will, justifiably perhaps, see you as their oppressor.

One of them walked in through the front door of our second-floor flat. That happened because neither of us ever developed the habit of locking doors; in fact, I never locked doors till I left Houston. I must have been aware of crime; I was aware of crime. She must have had the same blind spot. This actually is a very good proof of McLean's Triune Brain theory. My thinking center knew I should lock the door. But it wasn't my thinking center that actually turned the lock; it was my habituated moving center that did that. This center had never made locking doors part of its routine.

The bad part of the theory is that our three brains take turns being in charge and communicate poorly with each other. So this young fellow, in the predawn, who did not know us, walked in, stood in our bedroom doorway, flicked on the light that revealed our snoring and said, "Don't move or I will blow your head off."

He had my wife get up and bring him cash and rings, including my gold Rice ring. I was face down feeling helpless, but every time I began to turn, he noticed and threatened more fiercely. He then half-heartedly and awkwardly raped my wife before leaving with his small heist.

Soon she moved out near her job in the suburbs, and I stayed in the flat. We divorced a year later. She was a wonderful person, and my memories of what led to our breakup are a kind of nightmare with dark fogs hiding reality.

Forty-eight years later, I corresponded with her, and was so glad to find out she met the loving partner she deserved. She never had children. After she left, I met women who invited me into the Black culture and into their beds. There were lunches at a soul food restaurant, Black Russians on a lazy Sunday morning, and standing in line to see the slain body of Black Panther, Fred Hampton. There was even late-night dancing at all-Black nightclubs, my friends and I excited to be on the same floor with Bob Love, the six-foot-eight-inch star of the Bulls. He was easy to spot even in the darkened room with all the other moving bodies.

Most of my immersion, however, came from my work at Hughes Elementary and the college courses we were taking. I had my misadventures and ultimately the 35 hours of credit towards a master's that I earned came to nothing. I was not a strong teacher candidate. While I was smart and creative, I was not attentive enough to the initial misbehavior that leads to larger problems. Nor was organization my strong suit. I got a lot of support from the several Black teachers I worked with, and quickly saw what great teachers they were. I came to admire them greatly.

Ultimately, the Black men let me know that they weren't as glad to have me in their community as the women were. When the school year ended so did my portal into Black culture. I am so lucky to have experienced this. Before Teacher Corps, I knew only southern stereotypes and Lightnin' Hopkins. Now after being allowed to participate with Black people in an intimate way, I had come to sense and feel our common humanity, that, somehow, we were the same. Too bad about the larger impersonal world.

Meanwhile, as I worked hard in Teacher Corps for minimal pay, I assumed that my draft board would go along with the deferment. As it turned out 72 of the 75 young men in the Chicago Teacher Corps were deferred. At some date around Christmas 1968, I received a letter from Selective Service System informing me that upon the end of the school year, I would revert to 1-A status. Poof, there went my master's degree, my teacher certification, and likely my freedom. I had three avenues of action: I appealed and was denied, I applied for conscientious objector status and was denied, and I researched going to Canada as Nathan did but without Feynman's endorsement.

As summer was arriving, I got another letter telling me to report for my pre-induction physical in Chicago. On that eventful day, I joined hundreds of young men, Black, White, and Latino in a queue that wound through desks, tables, examining stations, benches, and the soldiers who reminded us of the fate that awaited us. After being measured, the paper I clutched in my hand showed my height as 6 feet 1 inch, weight as 139 pounds, and build as medium. My next in line comrade was 5 feet 7 inches, 185 pounds and also medium build. Make of that what you will. It also noted that I had claimed to have a hernia, but after examination, it was stamped, stamped mind you, "Varicocele, left side, not large or dangerous." So I did pass the physical and would receive my induction notice.

OBITUARY

Jamie Bray, pioneering ex-official

By ALLAN TURNER
HOUSTON CHRONICLE

Jamie H. Bray, a former Harris County commissioner who promoted racial diversity in government and was instrumental in the creation of the Armand Bayou Nature Center, died Friday. He was 79.

Bray, who was a state representative from 1969 to 1970 and Precinct 2 commissioner from 1971 to 1974, died of respiratory distress, family members said.

BRAY Bray was the first commissioner to employ an African-American in an executive role.

He was honored in the Pasadena Hall of Fame for his work in saving the 2,500-acre Armand Bayou property from real estate development.

"He was a friend to everyone," said his granddaughter, Sauletta Wilson.

She called him a fair man

Please see **BRAY**, *Page B5*

Fleeing to Canada was not something I wanted to do or was that easy, but by this time I had my dander up about this horrible war they were compelling me to attend. Then entered a man who would make that decision unnecessary. He was a neighbor and friend of my parents. His name was Jamie Bray, and he and I had once had a discussion about the war. He now had been elected to the Texas State Senate, my mother had spoken to him, and he knew the chairman of my draft board. Thanks to his intervention I received a letter offering a "courtesy" hearing on my application for conscientious objector status. My mother had not spoken of this, so it came as a complete and welcome surprise to me.

Since moving to Chicago, I had not seen my parents so you would think I would remember thanking and loving my mother and father. Unfortunately, I have no memory of any of that; I don't remember the flight, being picked up, meals, my younger siblings, nothing. I wonder what went wrong. How deeply was I distracted?

In those days the American Friends Society printed and distributed material on dealing with Selective Service. In my case, their advice was to assert that you would never take another human life regardless of the circumstance. In a sucky and submissive preamble to the hearing, I shaved my beard before flying down from Chicago. At the hearing, my witness, Father Pat, showed up with his elegant new beard and testified to my sincerity. Probably, the beards made no difference; it boiled down to what I would do if the commies broke into the house about to kill my wife and mother. I gave a more or less planned response and had only to endure their palpable scorn as they granted my C.O. status.

I have often reflected since on whether I was cowardly in avoiding the service accepted by other young men, even my friends, which they fulfilled with honor. Or was I honorable and principled for standing up to a monstrosity. Men do compare behavior, their own versus their friend, or versus the crowd. Then judgment enters, "Was I worse than my brave brothers, or was I better than them?" Some questions perhaps should not be asked in the presence of others with a stake in the game. Fortunately for my ego, years later in a very tense situation my righteous anger boiled away my concern I that I was a coward.

My C.O. status required me to complete two years of community service at a satisfactory employer. I was not allowed to continue teaching. One of my Chicago friends had an aunt in charge of nursing at the Chicago Neuropsychiatric Institute near Cook County hospital. NPI was a teaching hospital for resident psychiatrists and psychologists with an inpatient ward for up to 25 mentally ill, often psychotic, individuals. They came from the community, mostly got free care, and were selected based on providing a variety of possible mental illnesses for the resident to learn from. It was a very interesting place to work. I was an orderly, and before leaving, became a Psychiatric Nursing Technician and part of the nursing milieu. Of course, before praising NPI, I have to wonder how much of the success there may have been due to new psychotropic drugs. All I knew about their treatment was to interact based on the residents' suggestions, organize crafts and ward activities, empathize and send goodwill to the patients, help them with discipline when necessary, and always be ready to tackle a psychotic outbreak. The psychiatrists who ran the place were like gods, but even a peon such as myself got to know the residents who needed help to fulfill their treatment plan.

There were a lot of pretty nurses working there as well, and after my marriage breakup, I got to know them much better than the doctors. I continued to have good friends who were black at NPI, but the nurses I dated were white. In 1972 after we had shacked up for a couple of years, I married one of them, Marsha Lynn Heasley. She became the mother of our four children and a willing partner in a life venture outside the bounds of normal.

July 1, 1972: It looks as if Marsha is paired with best man, Vince. John Laue is next to Vince and Mike Polak has the Afro. My sister, Annie, is second from right, and Marsha's sisters make up the rest of the bridesmaids. Marsha's brother is on the far left.

I completed my two years of alternative service in 1971. The job at NPI, working among young smart people, was intriguing. With money my father had given me, I began work towards a master's in Social Work, MSW. I lasted only one full quarter. I witnessed electroshock therapy, shoddy treatment, and above all a prioritizing of getting paid. It was nothing like what I had seen at progressive NPI. Anyway, I realized that the MSW path was something I did not want to follow, at all. My former direct supervisor at NPI had divorced her husband, a semi-wealthy Jewish man who lived near the University of Chicago. That part of Chicago was an island of wealth and integrated intelligentsia amidst the impoverished Southside. She passed on to me that he needed a manager for his plant store in the heart of this neighborhood. Thus, began my career in horticulture.

I worked there a year. The owner rarely clerked at his store, but rather searched for beautiful and uncommon plants. I devoured the store copy of "Exotica," the encyclopedia of plants and their families.

I remember in particular the beautiful lettuce leaf aralia. The job was captivating, and I learned a lot about many plants, but surely this was not a career? Odd that it turned out to be so. In July 1972, Marsha and I got married, Mother and Dad offered us the use of their motorhome for a honeymoon, and we took off on a six-week trip over the western mountains to California. On the way up to see friends at a resort near Pagosa Springs, Colorado, we drove through Santa Fe, then, up and over Wolf Creek Pass.

New Mexico, the Land of Enchantment. We were enchanted.

Castor beans grow big from seeds. They are poisonous, but nobody wants to eat it or its seeds. Those are pressed for castor oil, one of the most healthful substances there is.

This is some form of Epostoa, the old men and women cacti. This is one of the few plants I did not propagate. I bought small ones, knowing I could grow and sell them for much more. Behind it are two furry kalanchoes and to the left a lovely aloe in bloom.

8

Plants Alive

October, 1972, after the motorhome honeymoon trip, Marsha and I drove back to New Mexico and rented a house in Espanola, twenty-five miles north of Santa Fe and forty-five south of Taos. From our front yard, we looked out over the Espanola Valley and up to the magnificence of Truchas Peak. Marsha took a job at the small rural hospital near Embudo, the funnel, north of Espanola. My father at this time was anxious to put up collateral for a plant business. It was an exciting time of beauty and expansion. We began at my parents' home on Lake Livingston, Texas, and used the motorhome as a panel truck. In nearby Conroe was a wholesale grower of bromeliads. We spent $500 and filled the vehicle with these unique semitropical creations of nature propagated by men. We knew nobody else in New Mexico had them. Then we quickly went on to Santa Fe, where only one block off the Plaza, catty-cornered to the Inn at the End of the Trail, in a trendy building named El Centro Mall, we opened our charming shop with the purloined name of Plants Alive.

Shortly thereafter, my dad and I drove to Albuquerque and bought a 1973 Ford Econoline van for $3,388, what a great vehicle. It was a three-quarter-ton cargo van, half window and half panel, soon emblazoned with Plants Alive, Espanola, NM, 505-473-5602. Over the years it carried, plants, furniture, family, firewood, and once, even manure.

By our grand opening in March, 1973, our bromeliads were augmented by a large collection of cacti and succulents from a retiring nursery woman in Albuquerque. I found an excellent Albuquerque wholesale nursery, and Marsha found a 7,500-square-foot wholesale greenhouse near Glorieta that offered beautiful flowering annuals and baskets. Canoncitos Greenhouse was owned by Helen Wilson, a retired professor. Her husband, Bob Wilson, had learned the trade at Payne's Nursery, where I worked later. Just starting with a greenhouse, I learned a great deal from Bob about the nursery trade, the most useful of this being how to construct a potting bench and mix the ingredients for your soil. Bob had far surpassed the Paynes. He was an artistic genius with plants, but he only began a year before us and later had to leave the trade so he never reached his full horticultural potential. What he did grow to perfection were blooming annuals and that was just what we needed. Helen was charming, but overweight and in her fifties; Bob was a normal-looking guy, 30 years old. Her 320 acres were at the mouth of a small but lovely canyon that gathered only an intermittent creek in its uplifted arms, hence its name Canoncitos. We went for walks with Bob up canyon, and he was constantly picking up arrowheads, then tossing them back down. Bob Wilson became our best friend and a comfort for Marsha later when my pursuit of an "open marriage" became apparent to her. Bob personified a kind of magician, someone who worked and spoke in a manner that included "spirit," or is it magic or mystery? This idea of something better permeated Northern New Mexico in those days; we lived indeed in the land of enchantment. When we bought from Bob, he carried flats of flowers down to our van held high in the air above his head, one cradled in each hand, perhaps smoking a cigarette at the same time.

John Denver came to Santa Fe. I don't remember exactly when, but we already had lots of friends who attended with us. "He was born in the summer of his twenty-seventh year, coming home to a place he'd never been before." That was me, only in New Mexico, better than Colorado. It was a very romantic time. At Christmas, 1973, Marsha and I conceived our first child.

During her pregnancy, I had an affair with a nineteen-year-old from the shop downstairs, romantic time indeed. The twenty-two-year-old from the ice cream shop liked me as well. These young women posed for me a great conflict. I found that I was unable to make physical love with a woman and not "fall in love" with them. But falling in love did not like to be shared. Life didn't allow that.

After Marsha discerned the truth, she allowed us to have a semblance of love and family life, but things were never really the same. Our daughter Arena Marsh Mueller was born in September, 1974. Marsha, of course, was no longer working, so our income needs fell upon the shop. The shop was doing quite well for the first year. It was returning a profit, albeit not enough for the three of us. I had also continued borrowing and built a 26-foot-by-48-foot greenhouse on the rural property of a new friend. It soon turned out he was about to have his land repossessed, so I had to buy it for $5,200; what a mess. I eventually got the money back and tore down and rebuilt the greenhouse in another temporary spot.

With the greenhouse, we were able to grow plants cheaply and sell them in the store, earning excellent profit. While stocking and displaying beautiful plants for the store was enjoyable, it was at the greenhouse that I began gathering the tools of horticulture. That began with soils and propagation. Watering, I already largely understood. Although I would not have said it at the time, proper watering comes about as a result of the connection between the grower and his plants. To any sensitive individual the plant will convey its needs.

In horticulture, soils are of the utmost importance, and so any study must begin there. For me as a practitioner, it was more difficult to access knowledge than had I been a horticulture student at a state university. At the same time, it was more relevant and important. Honestly, I do not know how I researched anything before the internet, but I somehow heard of the John Innes mix from English horticulture, which was compost oriented, and the Cornell Mix, from the noted School of Agriculture at Cornell University in New York. In the trade, the Cornell recommendation was called a peat-lite mix and is largely Canadian sphagnum peat and perlite. Interestingly, developed in the

30's and 40's, the Innes mix relied on manures to provide nitrogen and they were much heavier in a pot and were watered much less often than the Cornell mix, which developed in the 50's after the power and convenience of synthetic fertilizers had become available.

In the Innes mix, the composted manure would supply minerals and microorganisms missing from the peat. Ground minerals could be and were added to the Cornell mix. Practically no nurserymen I knew of, certainly not Bob who was a peat, perlite, lots of fertilizer grower, worried about this element, and most growers I visited grew only in peat and perlite, or peat and sand. They were confident that their synthetic fertilizers supplied all the plant's needs. This thinking was already not for me.

New Mexico's enchantment soon led me to a man who knew. His name was Dan Boardman, who in his previous life had been a successful civil engineer. He had "dropped out" and was living with his wife and two boys on a couple of irrigated acres in a classic large adobe home in Chimayo. This was a town well known for its Spanish weavers seven miles up canyon from Espanola. Here he kept a milk cow, hog, and poultry. He had an excellent garden and cooked and heated with firewood. Dan's wife, Harriet, supported the effort teaching school, and Marsha and I loved her very much as well, though we would not have used that word at the time, especially in regards to Dan, who had an overwhelming presence.

He was serious and passionate about the environment, of course, and he lent me four of the great works of agriculture, *The Soil and Health, Humus and the Farmer, Farmers of Forty Centuries,* and the 1938 USDA yearbook, *Soils and Men.* Dan had an admiring group of a dozen or so young hippie, would-be farmers and also local Spanish admirers as well. He was eighteen years older than I, and Harriet was six or seven years older than he.

From local leads, I was able to fork manure from stables and corrals, filling my 1951 GMC pickup till the springs went flat with the weight of horse, cow, buffalo, or goat manure. The heating pile that resulted back at the greenhouse opened me to a world of life and a feeling of the energy of creation. After one turn and two months, I screened it into my potting soil. Actually, for my bedding plants that was all I used for soil, screened compost. For flowers, hanging baskets and bigger plants, I added peat and perlite mostly to lessen the soil weight. After this beginning, I continued composting and mixing my own soil over the next 45 years. I still bought peat moss and procured perlite, but I never mixed any potting soil without an element of soil life incorporated. The soil created a gap between me and every other horticulturist I met, none of whom had ever pitchforked manure out of a stable and made compost.

Composted manure included in a potting soil made it a repository of life and life energies. This soil did not compromise with typical synthetic nursery soils, but distinguished itself as the champion of nature, opposed to the beautiful lifeless blah of industrial production.

Marsha and I were out in nature every opportunity we had. Near were Santa Clara Canyon, Bandelier National Monument, the Rio Grande, and the high Pecos country. There were meals cooked over open fires with the wood gathered at hand, sleeping in tents under the stars and bright moon. We thought everything would keep getting better.

But unnoticed by us at the time, an event on October, 19, 1973, would see the end of the romance. On that day the twelve members of OPEC approved an oil embargo on the West. By December, Plants Alive's first Christmas, we, every American, knew we were in trouble. Our business and a very big chunk of Santa Fe depended on tourist spending. There was no commercial air service and the train depot at Lamy was 20 miles away. Everyone came by car, and gasoline prices quadrupled within six months. There were gasless Sundays and 8-gallon maximum fills on the interstate stations, just enough to keep you from running out of gas.

The fact was we couldn't sell anything to someone without a car. Store revenues dropped like a rock, and our second year was 20 percent below the first which had begun to fall at the end. We closed the store in March, 1975, and for the rest of that year and part of the next, we remained in business with just the greenhouse at a decent location on U.S. Highway 285 coming into Espanola from Santa Fe. We were rolling the dice on a spring bedding plant crop in an agricultural community.

Over the winter, my friend Jon Somers from college helped me rebuild the greenhouse on the front of our rented adobe. By earliest spring the 26-by-48 greenhouse was attached to the porch. It was now glazed with high quality fiberglass panels instead of the Monsanto 602 polyethylene. I remember it as sublime to be in this old and lovely home with the greenhouse open to the porch.

The spring 1975 crop was lovingly grown and very successful. I grew 14,000 organic vegetable and flower starts in pots I made with paper mache, Dan's fabulous idea, of course. Every inch of space had a sellable plant. We sold out in May and June. There were also trips to Bob Wilson's for many of his beautiful flowers. In May, we sold $4,300, by far the most our business ever achieved. We followed May's sales with $3,200 in June. That money helped immensely, but after we paid debts and expenses, we had no ability to buy inventory. Nor did we have the store to sell it in. House plants, succulents and hanging baskets that had sold well in Santa Fe weren't as well received by the rural Espanola that had just bought our tomato and pepper plants. Our business needed to be a complete nursery, not just a greenhouse.

In those days I knew little about shrubs, trees, or perennials, much less how to propagate them. So both the money and the know-how were lacking, and spring of 1975 turned out to be the high-water mark of Plants Alive, Version One. At the same time, we worked toward self-reliance amidst the stunning natural beauty of Northern New Mexico and its awesome cultures.

I was very fortunate my last couple of years in New Mexico to rapidly learn more horticulture. I gained a lot of confidence and time proved that my knowledge, while not unique or even original, was a rare and valuable thing. At the same time from the homestead came pork, eggs, canned and frozen vegetables and home baked bread, all of which supported us against the slide into poverty. Doing and learning these agricultural and food-prep skills was more meaningful and useful than making a B at Rice, and so brought a sense of pride and self-reliance. Those valuable skills helped with the groceries but did not pay the rent.

We entered the planting season with fewer resources and the sales season with less to sell. In the fall of 1976, we were compelled to move out and take our greenhouse with us, another huge effort with no income.

My brother, Chris, moved back from California to live with us, and he helped slaughter, dress and freeze about 30 chickens. Since we also took the meat and lard from our hog, including two beautiful hams that we kept in brine in two big crocks, we moved with plenty of food.

This blue rosette is an Echeverria.

This Senecio is free flowering with beautiful orange flowers.

This is our original "dying" redbud the last spring I saw it. After I left a summer squall tore it apart. I am sure it had been rotten for years.

9

Santa Cruz River Valley

Espanola sits at the bottom of the Rio Grande's only floodplain in Northern New Mexico, well named the Espanola Valley. The city straddles the river with only a single bridge to connect east with west. The valley is about 30 miles long north to south, and is met from both east and west by smaller streams that each have their own little floodplains and acres of irrigated land. Land in any of these valleys with water rights is very valuable and most plots are small. Outstanding crops of New Mexico peppers and corn grow in the bigger plots and gardens provided all our favorites. The soil where I gardened must have been fertile because I did nothing special. I think they were my best vegetable gardens.

Large clumps of asparagus grew along the roadways, ditches and lanes. I literally filled a grocery bag in less than an hour. There were many small orchards, and apples were a reliable crop. We were there when the once-in-ten-year crop of cherries came in, and Marsha and her friends gathered to pick them. It was the same with apricots which bore heavily every few years, so good it all was.

Some of the small valleys were owned by Pueblo people. On the east were Taos and Nambe, on the west San Juan. The pueblos were mostly closed, and the majority of my contact was with the Spanish culture long in place. The Spanish seemed to own most of the Espanola Valley, and the big towns were Embudo, Dixon, Lyden, Abiquiu, Velarde, Alcalde, Santa Cruz, Chimayo, and Hernandez. I somehow got to know a lot of them well. The face that Espanola offered to the world and to myself was Spanish. The Rio Santa Cruz became the entitling tributary of this chapter because I lived in its valley and fed off its waters which flowed from the western flanks of Truchas into the Rio Grande just south of town.

More of the Indian culture lay in the massive Jemez Range to the west; actually, Jemez is one single extinct volcano. The walls or plateaus of this upland were created by the massive, thudden fall of pumice, ash, and scoria all around the ejecting caldera. This rim of mountains and plateaus were soft boned, unlike the Sangre de Cristo Mountains, and deeply cut. The mesa tops of 7,000 feet were clothed in pinyon pine. Pinyon could grow pretty big, 50 feet maybe, but most were smaller. They didn't seem to offer much in the way of pinyon nuts while I was there, but they were the best wood for the stove or fireplace. Perched on the east face of the mesa was Los Alamos and its bedroom, White Rock. These places would not have looked too strange back in Houston, but here they are bizarre intrusions on the mountain.

On the same eastern slopes of the Jemez were two Pueblo ruins, Puye Cliffs and Bandelier. To me, Bandelier was mind blowing. There could be no better place from which to approach an understanding of this historic culture. It helped greatly to have read *The Delight Makers* by Bandelier himself.

The canyon was deep and narrow but wide enough for sunlight and habitation. Water flows all year long, at least in our era. The south-facing wall of the canyon held the classic adobe structures gathering sunlight in the lee of the cold north winds. Bandelier is not as extensive as Mesa Verde, but it has everything, and at that time was more accessible.

On a Christmas Eve, I was with Marsha, her sister and our brother-in-law on an ancient, flood-cut ledge on that wall. I remember it as perhaps 100 feet by 50 feet with an overhanging roof ten or twenty feet tall, scoured by some ancient flood. We had climbed up ladders and steps constructed by the national park to this ledge far above the canyon floor.

In the middle of the ledge, at our feet was a large kiva, and there was a ladder down into its mysterious ether. I wasn't comfortable down there. I am sure the kiva was 12 feet deep and 25 feet across. The Puebloan builders must have excavated the tufa to create it.

As nightfall approached, we sat outside on the ledge facing south over the canyon. The tops of the ponderosa pine growing up from the canyon floor were at our eye level, and at that moment it began snowing. It was like being in a glass holiday bowl someone had just shaken. We were the only ones there, everything was still with the love of nature, and that moment remains like a still photograph in my mind and in Marsha's as well.

On the east side of the river just at the south end of the Espanola Valley and near the beginning of White Rock Canyon sits Otowi Mesa. Anglos called it Black Mesa and it was very dark. It lay on San Juan Pueblo land and was said to be sacred and not to be trespassed upon. I was then and remain today somewhat confused by the indigenous meaning of the word sacred. I thought it was like, "All life: sky, soil, water is sacred;" then I thought if everything is sacred then I too must be. I would never defile Otowi Mesa, never leave a candy wrapper or a cigarette butt; I would, however, gladly climb Otowi Mesa, and I did. There were few, if any fences, but there was still a walk in over the eroded floodplain of the rio. I had climbed a mesa before, when I was sixteen, with my father. That mesa was almost as big. It sat above the canyons falling down into the Canadian River basin near Canadian, Texas. At that age I missed a lot, but the view from the top seemed to turn me into a whole new person and I wanted to be "out here, in this."

Otowi Mesa was anticlimactic. Its geology, its water shedding channels, its flora and fauna was all fascinating, but on a hot day, this 28-year-old was not a student of any of that. Oddly enough, though the view from the top must have been special, I have no memory of it. Did the spirits take it away?

But just as we got back to the van, we heard a distant goose call. We watched as a series of flying-V's came into sight. I recall seven or more of the V's forming a kind of second layer of them, V-squared, I guess, like WWII bombers. The formation was flying at the same altitude and quite high up. They had just flown over the much higher plateau west of Santa Fe.

From where we stood the Rio Grande was about to enter White Rock Canyon cutting straight through this obstacle. The plateau just traversed by the geese was 6,300 feet and we were 900 feet lower. Suddenly with great noise and cacophony, the geese began to resemble an assembly of gnats by a light, or a school of darting fish. The whole swirling mass of geese lost its forward momentum and began to twirl down towards us. Then after a harrowing minute or more the geese reorganized themselves, perhaps with different "point men," and sped north again up the valley. Looking due east from most anywhere in Espanola one saw the north and south peaks of Truchas, the grandest in the Sangre de Cristo Range. The peaks were bare. In the summer they presented with green flanks and silver-grey rocky tops. In the winter the peaks were white, and the green of the flanks was misted over. I climbed lesser peaks, but not these. Their western slopes were deeply cut and eroded in the high country, but further down the land leveled and was nourished by the mineral rich water that coursed down the canyons. From their earliest arrival, beavers and humans began capturing the water and holding it in pools. Humans went a step further than the beavers and diverted water from the pools into hand-dug canals. These canals were cut to the outer edges of the different flood plains held within the little rivers like the Santa Cruz.

Some of these canals were built in the Seventeenth Century by the invading Spanish conquistadores and their accompanying families. Other canals in neighboring valleys had been created by the Pueblo people and were even older. Topography would eventually force the ditch to empty back into the river bottom, but a little bit downstream another check dam would take the water to the rim of another little floodplain. Each of these new ditches sponsored the fields and homes of a different tiny community: Sombrillo, San Pedro, Chimayo, Cuarteles, La Mesilla, Truchas. In 1937, the Santa Cruz Dam was built, and the reservoir captures some of the snow melt from the peaks and holds it to assure water later in the season.

Every landowner with water rights is allowed to draw water for their gardens, trees, and pastures. In return, everyone is required to work cleaning the ditch in the early spring. All this water movement is regulated by the mayordomos. In dry years, they wisely direct more water to the best farmers. My guru, Dan Boardman, was one of these. I was not. But I did work on the ditch with my mostly Spanish neighbors, and I did thrill to the flow of water creeping down my rows of vegetables. I was still a kid back then, but now I know that this water not only carried the minerals and memory of Truchas but existed on a spiritual plane as well as the physical one. I lived five years in the valley and had my most productive gardens there.

Our first garden was very successful, I'm sure the best I ever had. I now only remember a whole lot of fresh peas, 54 pounds of carrots and 75 pounds of turnips. I raised and slaughtered a hog that I had bought for $25 as a spring wiener pig. We raised about 30 barred rock chickens as well. Both species had decent pens. The following year I added wheat, triticale, rye, and cannabis to the garden. I grew the grains in the manner described by F.H. King in *Farmers of Forty Centuries*. That involved planting in well-spaced clumps so that there could be cultivation or intercropping. This importantly meant that harvest could be done with a sickle. The grains grew rapidly in the spring and hid the cannabis plants inside the grassy perimeter. I actually harvested it all, over 200 pounds from my small plot. I made a threshing tool, threshed and winnowed as well. Most of the grain, I fed to the chickens. My triticale fascinated me; the heads were 50 percent bigger than the wheat, which was excellent, and my chickens adored it.

So while my grain crop management came from the Chinese, my cannabis culture was from the local Spanish tradition. Seed was planted in the garden in November as if they had fallen from a mature plant frozen hard by the early bites of winter; it was meant to lie dormant in the ground until it knew when to germinate. In March, the seedlings emerged just peeping one-quarter inch above the soil, but with a head the size of a quarter, really unlike the same plant grown in a windowsill, or even a greenhouse. They were covered by a late spring snowfall, but on they came. Fifty seeds yielded 16 plants. Nine became females. All grew to seven feet tall. Following the guidance I was getting from Dan Boardman, I pruned them as was customary in the valley. Actually, the plants were pruned twice, each taking out the center while leaving the tips. The shocks to the plant, suddenly losing 75 percent of its leaves, induced DNA multiplication or polyploidy, brought more resin and higher THC. That is to say the cannabis plant when shocked or stimulated in some way can add new DNA to every cell, changing its own genetics. Often the strands are doubled but with different genes on the new strands. The sex often changes and some plants become hermaphroditic. Resin is for cannibis a protective substance against drought, certainly, but also suffusing the seed for cold protection and must have other values I cannot imagine. The pruning also multiplied the flowering tips and formed a willowy plant that had less stem tissue.

Turkey or Chinese rhubarb is one of the four components of the Essiac cancer treatment formula. I got it to sprout but could not get this plant to survive the hot Oklahoma summers.

The snowy cactus is Opuntia vistita. All the flapjack cacti are in the genus Opuntia.

Mullein, lamb's quarters, peach tree behind: I have eaten a lot of lamb's quarters. I enjoyed it but also remember the fur on my teeth afterwards. It is a quinoa and seeds must have food value, but it is far from edible quinoa. It may be a good bird food. Perhaps, it is worth it to collect lamb's quarter seeds just to keep them from all falling to the ground and sprouting next year.

10

Arena

In a beautiful moment fully in accord with Romantic New Mexico, our first child was conceived in front of a warm Christmastime fire. She was born September 21, 1974. Outward aspects then were not auspicious. The recession at that time was palpable, the store was treading water, and we were living off loaned money. My parents had moved up from Texas, my siblings with them, of course. The twin boys were 14 years old and already getting into trouble. My sister was 17 and also rebelling in her own way. I was having an affair, which though presumably hidden, still managed to sour everything. Marsha and my mother did not get along, the boys were in hot water, the store was not enough, and we already knew that it would be dropped when the lease ran out in March of 1975. So before too long my parents left Santa Fe for California.

Arena Marsh Mueller, on the first day of fall 1974, in Virgo, but near its cusp with Libra, was pulled out of her mother's womb with forceps. I was standing there astonished at the degree of force the doctor exerted to extract my child. Marsha had been given a saddle block so as not to feel the pain. She was only about five feet tall and a hundred pounds before her pregnancy, and Arena was waiting to be born "sunny side up," or in a sense, upside down. Marsha was unable to push the seven-pound-eight-ounce baby out. This was the kind of condition that in times past either killed the mother or required cutting up the baby.

Arena

I realized that from my vantage, my first view of Arena would be her face, and I stared in alarm as her eyes and nose failed to appear. This was a shock that has stayed in my consciousness from her first day. It passed within an instant when eyes did appear. There was a lot of stretching and temporary distortion of her head from the forceps. She was fine, more than fine. Arena got a lot of love from both parents, grandparents, uncles, aunts and friends. Being a first-time father brings a kind of shock wave that is hard to understand. What used to be free time is now time holding the baby, wanting to hold the baby, changing, later feeding, always caring for. I continued to work at the store in Santa Fe, so I recall that my sexual dalliances probably continued for a while, but not for long.

Being a father put that in a new perspective. So over the next three years while Arena went from baby to toddler to child, I was there as I should have been. Marsha, aggrieved, became soul buddies with Bob Wilson, who had divorced his wife and left the greenhouse. Bob as an artist continued to embody the enchantment of New Mexico. He gifted Marsha with his beautiful paintings of New Mexico, mother and child, and plant devas. They spent a lot of time together, and I felt the sting of jealousy and a fear of abandonment. Meanwhile, my attempt to save Plants Alive became like a rear guard action in a losing war that began before Arena's birth and was waged to keep us in New Mexico.

Sedum morganium or burros tail. It's easy to propagate, but hard to grow big.

This Star Jasmine rooted through its pot into the ground then grew 12 feet tall and 8 feet wide, blooming like this every late winter and filling the greenhouse with fragrance.

The yellow Clivia is just as pretty as the orange. I grew some from seed but after four years they were still small, just about the slowest plant I ever grew.

An oak leaf hydrangea with a false Solomon's seal neighbor. Horseradish is hidden under the big pine. I cut my own water barrels. I often didn't use them that much. Mosquitoes? No, I netted guppies from our aquarium inside and stocked the barrels in summer.

11

Chicken Little

In the spring, I bought 36 straight run Barred Rock chicks. These are a popular, large breed, considered to be dual purpose, eggs and slaughter. It was easy to build cages underneath the greenhouse benches, and there it was safe and warm for them and they thrived. The birds were as advertised except for two remarkable exceptions. One was a pure jungle fowl with all the colors, black, vivid green, red, gold, and russet red. He shimmered all over. He was the only chicken I ever named, "Golden Blackie."

The other was a leghorn cockerel. Leghorn is the smallest commercial breed. The hens are thrifty and lay lots of eggs, but the roosters are worthless. Before I built the larger outdoor pen for the chickens, the little cockerels were already fighting for dominance. One large, barred rock was very dominant, but I saw him back down to Golden Blackie, who was much smaller. The outside pen was a six-foot-high fence around a good-sized Colorado blue spruce. From the very first day outside the two little roosters flew the coop and spent their days in complete freedom, and at night they roosted on top of the same fence they had surmounted. Meanwhile, the larger birds remained trapped inside and the big rooster reclaimed number one for that society. At night these birds mounted the tree in their pecking order. They would not share a branch. Roosters up first and highest, the hens below, and at the very bottom was the pecked-out hen and her sisters above her in obvious pecking order, with the one clean hen at the top of their society. If a coyote came, the bottom bird would be the victim.

Golden Blackie spent his youth outside in the midst of an acre and a half of alfalfa and other good herbs. However, the leghorn hung around the greenhouse becoming a pest. Later I found his depredation. He had eaten a whole flat of cabbage plants worth $40!

We were leaving for a couple of days, and I decided to force all the chickens back into the greenhouse pen. I managed to snare the two escapees off the fence to join the others already inside. But the instant I dropped the jungle fowl in, an electricity ran through the flock. From ten feet away, the big rooster with feathers flared announced his challenge. The two now full-sized birds rushed together and GB was flipped backwards by his stronger opponent; then twice more with the same result. After the third knockdown when GB stood up and advanced, the big barred rock turned and ran. Pedigree re-established.

Months later, my 16-year-old brother had come to stay with us. We set out to slaughter all the roosters so as to go through the winter with just Golden Blackie and the hens. There were about twenty large roosters to kill, pluck, gut and freeze. We were nearly done when I came upon the leghorn rooster trapped in the pen. The killing lust was upon me. I was going to snatch this worthless bird and wring its neck. Racing after him down low I dived for his legs and cracked my noggin on a low branch of the spruce. Bleeding and dazed, I left it with my brother to finish the job. Then all the dressed and wrapped birds were in the freezer.

I should mention that these carcasses all had tiny breasts and huge stringy thighs and legs. They were stewing birds, not for tender frying. Nonetheless, when our friend Vince came through for a visit, he asked to take one out to his sister in California. She had purchased a new chicken roasting oven that was all the rage out there. Vince wanted to show off a real naturally raised bird. He left in a few days with his frozen bird packed in his luggage. Not only was it the scrawny leghorn, but Chris had done a dismal job of cleaning the worthless carcass. The feet were still on and many feathers remained as well. And so this bane of my poultry operation ended his days not in the stomach of a coyote, which should have been his due, but in a dustbin in Los Angeles, California.

In a final epitaph, snows came after the adventure. I could not find my glasses. They showed up under the spruce four months later when the snow had melted.

This Monstera leaf looks pretty ominous. In the tropics it grows even larger and is the source of breadfruit. This one grew through its pot into the ground, but was in a crowded place so never got fully developed. It was very happy in winter temps that touched freezing.

Not sure why this is called peanut cactus. It has striking form and color. I believe it is Mammalaria elongata.

I love this deciduous azalea, flame azalea. That's a chunk of burdock next to it. No, I did not want it there. Burdock is a powerful medicine plant, and I started with one or two. Soon there were a hundred, many a hundred feet from where they started. I didn't mind them and they were beautiful in a unique way.

12

1977

I took my second job in horticulture at Payne's Nursery in Santa Fe and encountered the world of pesticides. I did not like Lynn Payne, and after a while I got fired. That came as a result of a conversation I had with an important customer about whether the pesticide Lynn had sold her was safe for her children and pets. It had been applied to plants around the swimming pool no less, and wasn't safe. The Payne's did lots of dangerous and illegal pesticide applications. I was 31 years old, but nowhere near "mature" enough to function in a world that didn't care how smart or environmentally conscious I was.

At this low moment in my life, our friend Harriet Boardman sent me to Red, the owner of Red's Steakhouse, a dance hall and drive-through liquor on the strip in Espanola. Red was hiring a weekend security guard and I was it. The $60 per week was not enough to live on, but very welcome. I was surprised to find out that as part of my employment, I would become a deputy sheriff.

The sheriff of Rio Arriba County was Emilio Naranjo, de facto boss of the county. Emilio was undoubtedly getting kickbacks from Red and gladly deputized Red's candidates. Otherwise, the notion of me being a real deputy in that county is pure comedy. Our hippie circle compared Emilio Naranjo to Richard Daley.

So, for my first Friday night, Red had me come in early with my card and wearing blue or black slacks and T-shirt. He outfitted me in a blue epauletted shirt and a Sam Browne belt with .357 magnum pistol, baton, mace, and six-battery flashlight. Thus garbed, I was continually having to hitch up my pants. Red was 6 foot 5 inches tall and broad, a martial artist I imagine, so I, in my job watching out for stuff, was good, responding independently to anything probably bad. Red said if I saw any disturbance, I was to flick the flashlight beam in his eyes. When my first fight broke out the next night, Red was about fifty feet away but caught my signal. Being much closer, I piled in by jumping on the back of the aggressor. He shook my 140 pounds, plus equipment, like a terrier shakes a rat. Red arrived and dragged the both of us across twenty feet of dance floor, then shoved the miscreant out the door. I almost went out as well, but hit the door frame and bounced back inside.

Later around 3 a.m., I stood guard while Red counted his money. He paid me my tiny portion, poured me a drink, and said he was pleased with my prompt signal and enthusiastic engagement. He also showed me a common wrestling hold called a full nelson, which can control even a much stronger opponent. This was my first martial art; I know it to this day. With me applying, it should perhaps be called Art Marshall (king of pratfalls).

I worked for Red about six weekends altogether, and the rest of the time was mostly uneventful until my final Saturday, when suddenly, there was a catfight. The other dancers pulled away and a kind of ring was formed around the ladies. They had latched on to each other's forearms, bloody scratches involved. Red, I found out later, was outside dealing with some ruckus in the drive-in lane. So I pushed through the crowd to address the situation myself. The girls were very angry, pulling hair and wanting to scratch the other's face. Trying to pry them apart was worthless; they ignored me.

Then I quickly and effectively applied the full nelson to one of the combatants. She went limp in my arms and when the other, somehow startled or overbalanced, let go and pushed, my prisoner and I became overbalanced as well. Slowly at first, then faster, we toppled over backwards and my pistol, baton, flashlight and mace hit the ground an instant before my head bounced. My legs remained extended so it had the purity of full extension and for my noggin the full quarter perimeter of arc. My prisoner was unhurt as I completely cushioned her blow. After that, the moment of fury passed and comedy triumphed. The two girls even continued dancing with their respective friends. On that successful and creative, deputy sheriff note, I ended my career in law enforcement.

During that time, I did rebuild the greenhouse, hoping against realistic hope. I paid a backhoe operator to dig three-foot deep and wide aisles inside the structure.

It was the largest greenhouse I ever built. Eventually I would have figured out the necessity of cutting the fiberglass and framing in vents on the top ridge. In the fall of that year, 1977, our potter friend gave us $1,000 for the greenhouse and we used the money to return to Chicago, where Marsha's parents were anxious to help us start over.

Amaryllis are easy to keep going all year, and in winter, I would cover them with mulch under a bench in the greenhouse. All this foliage and flower grew back larger each year from the dormant plant.

How I have come to love mock orange. I hacked out an 18- inch-sprout bare root, took it down to my neighbor, and we planted it in a tiny bed in front of her house. It took off, and four years later hers was as beautiful as mine.

13

Hinsdale, Illinois

After a gloomy Thanksgiving, we loaded the van with Arena and all the possessions we could carry. It was heavily laden. Marsha was driving east on Interstate 40 after we had dropped down from Glorieta Pass to the high plains that extended all the way to Oklahoma City. She drove in a light windy snow and we were dropping down to the valley beneath in temperatures below freezing. On either side were wide highway swales that looked drivable. Marsha asked what she should do if we swerved. I'm sure she must have felt the van getting light under her feet. I replied if we slid off the road to keep her foot feeding some gas and steer for straight and then recovery, then keep it moving.

Within a minute we dived almost perpendicularly left. She did beautifully, the van was bounding in the air like a four-wheeler but kept its stability. She took a wide swing left, plenty of room; we were pushing right through the sage brush, plenty of gas on the turn, and it felt like a stagecoach as we pulled back onto I-40.

Driving much more slowly now, we passed a couple of spun out semis on our way into the next town, Tucumcari or perhaps Santa Rosa. I believe a lot of people know that this part of I-40 all the way through the Texas Panhandle is one dangerous highway when wind, snow and sleet come to visit. In calmer weather we visited with my grandmother in Canadian, Texas, the next day and then drove on north to Hinsdale. Our pretty blue van still had "Plants Alive Espanola, NM 473-5206" prominently hand painted on the side, so we didn't exactly fit in with our new neighborhood.

We moved into an apartment in November 1978, just a few blocks from the Heasley's lovely Hinsdale home. Marsha took a nursing position so that paid for our existence. Marsha's father wanted me to make a career for myself and asked me to read the book, *Dress for Success*. Actually, this is a useful and classic book for doing well in the world of business. I just hated regimentation. Back in the Teacher Corps days, I felt the same way when a Chicago principal had her students form ranks and march to their classes. Now I understand her wisdom, but in that time, I shrilly condemned it. Now I have to consider the arrogance I manifested, and how that cut me off from people, from life. Not that I can do anything about that now.

I soon did get a job that I liked.

My job was at a beautiful old glass greenhouse range in Elmhurst about a half-hour commute. After my first week, I was made head grower, so I had a lot of freedom and a bit more income than minimum wage.

Here I am at Morgan's Flowers in 1978. One of a very few pictures of me clean shaven.

About ten months later, I took a better-paying job at Morgan's flowers in Elgin, Illinois, 30 miles west of Hinsdale on the beautiful Fox River. Shortly after I started at Morgan's, Marsha asked me to leave. I was surprised. It was just that I thought I had been an upright husband and father since we left the enchantment of New Mexico.

So I found a little house to rent adjoining a swamp just above the Fox River. The ground at the back of the house sloped steeply down to the cat-tailed swamp which was the home to many red-winged blackbirds that filled the air with their powerful songs. Here I had the second of my "great bird experiences." Cautiously exploring the defile behind the house, I startled a large male pheasant. Of course, I was also startled. This big bird lofted itself, apparently straight up from its cover about six feet from my approach. The power, the flight pattern, and the noise were amazing. Right to the top of a 40-foot tree, I could see him clearly as he surveyed his field. Then with a jump he launched himself into a glide of 300 feet to the heart of the swamp scattering blackbirds as he hurtled down. I lived there about eight months and remember the place and the Fox River Valley with fondness. Arena also saw these pheasants and blackbirds; I wonder if she remembers.

During this time Arena, now five years old, was my most important friend. She spent every other weekend with me. I was, as they say, not in a good place, but I was still a good father. Arena and I walked to church Sunday mornings. I don't remember meeting anyone there, but I know everyone was sympathetic. She remembers spending the night and climbing on the iron stanchions of the front porch. From the beginning of her ability to talk, Arena had been bright, charming and well spoken. It was apparent that she was of very small stature, but we didn't think she would end up under five feet tall. Never mind, she had a big voice, once breaking a quiet moment with all seated at the Heasley's Easter table, saying, "Jesus Christ, there's a leaf in my salad!"

She also brought the carrot joke into our family, though I quickly appropriated it for my own. My relationship with my tiny daughter undoubtedly convinced Marsha to try again, and made possible the loving family I have today. I could say I owe Arena much of what I hold dear.

Arena was almost six years older than Graham, so she has memories of the years in Jemez, Flagstaff, and Mexico. She had a little bit of Montessori, first and second in public schools, and a bitter, little taste of Mexican primary for schooling, but most of her early education was at home with her mother and siblings. I joined in evenings and weekends when we were either reading or off on some nature activity. We lived in places where nature was nearby or engulfing. When we landed in Oklahoma, Arena started in seventh grade and went to the top of her class right away. Five years later, she was valedictorian of Midway High School. She had gotten ahead on her classes, and in her junior year she entered a free vo-tech program and earned a cosmetology license. She began her higher education at a junior college, transferred to Northeastern State University, our small but local and degree granting university, where she earned a BA in psychology. She paid for all of that education working as a stylist in Muskogee. Moving into independence, she became even more of a dynamo than she had been as a child.

She moved to Kansas City and got her master's degree, still cutting hair and earning over $30,000! Her PhD came from the Adler Institute in Chicago, where she was able to stay with her beloved grandparents. Since then she worked in Texas for a year or so, and for the first time since Elgin, Illinois, we were able to visit as father and adult daughter should visit.

Today, in 2020, she has a fulfilling career with the Veteran's Administration in Tulsa. She's married to Walter, one of my favorite sons-in-law, and has earned family recognition as someone with "good judgment" and is the de-facto queen of the Mueller's. I am so incredibly lucky that Arena and I are still close and see each other often.

First year with the greenhouse. Those are cucumbers heading up.

Unmown grass, sparse with wildflowers at first, but getting better with time. I mowed more or less once every three years. Trees I planted like this pine have grown well with no watering.

The tall, slender pear in the background is a clone of the T-Town Pear Trees. It's an oriental pear, known to its aficionados as the apple pear. I found a second pair of this cultivar on an abandoned Tahlequah home-site. I am guessing 19th-century settlers in Oklahoma were propagating and sharing this tree.

14

Old Greenhouses

The Elmhurst job was two suburbs north of Hinsdale. My employers were a son and his bride, taking over an old greenhouse range from his parents. It was 83,000 square feet of glass and was just being converted from cut flower production to container production. The Breiter range looked a lot like this one.

These structures were probably built in the 1920's. They mostly had wooden or abestos benches filled with Illinois topsoil which they steamed using their coal boilers. Heating was from hot water pipes under the benches. The universal crops back then were cut flowers, including mums, snapdragons, carnations, stock, and lilies and for 50 years, they had supplied Chicago florists. I am sure they existed on this model in all American cities. They had vents at the top which you can make out from the picture. They had no electric fans for ventilation, but accomplished that with natural air flow or convection. They were beautiful and relatively efficient structures, burning coal and using hot water pipes to pass heat from the boiler to the pipes under all the benches. These had a wonderful heat knocking sound to go along with a steamy tropical smell.

Sand's parents and Dixon

They were totally different from the all-plastic, forced-ventilation, space-heated greenhouses I had been in in New Mexico. I am so lucky I had a chance to work in a couple of these ranges because I think they are all gone now. My sister Annie and her new husband, Dixon, came to visit us.

Dixon was a juggling, flame-swallowing, unicyclist, circus troupe performer. Annie was the cheesecake, and they had a third fellow who also juggled and did tightrope stunts. They booked and performed at malls and town fairs.

In the bitter winter of 1978-79, just before Marsha asked me to leave, I began working as head grower at Morgan's Flowers. Three gay florists operated a 25,000-square-foot range where some cut flowers were still on the benches. I was their expert on container plants. I worked there just about a year. I had a satisfying little affair with another employee, smoked pot in the office, accidentally leaving the roach. Then I squabbled with the sales manager who overpriced everything, thus leaving my well-grown but perishable plants on the benches. I wasn't fired, but it was another sour ending. I quickly found a third Chicagoland job in October that paid my highest salary yet, $280 a week.

Then Marsha came back and we conceived Graham. I returned to Hinsdale to the same apartment I shared with Arena and now a swelling Marsha.

Wandering Jews can make lovely hanging baskets and they are easy to grow. This was probably December because my only pea crop is on the right climbing fast up the wall.

We had a lot of peonies and not nearly enough. The white and blue irises were my favorite and behind them an 8-foot-tall knock-out rose.

15

Graham

I remember, even though I was in a daze at the time, when Graham was conceived. After nine months of separation and minimal communication, Marsha came and saw me at my rental home in Elgin, Illinois, where I worked as head grower at Morgan's, a large florist and greenhouse operation. She was a nurse and living in Hinsdale about 35 miles away. Something had changed, I wasn't so bad, I was a good father. Arena was five, Marsha was 32, and she wanted more children.

I moved back in with Marsha and Arena and acquired a good little job with the Hinsdale Parks Department. Arena helped me with a little garden plot, and there were photos of us in the paper. I was able to not have any girls on the side with this pregnancy, so that made for a happy time. We were excited that the sonogram showed the child was a boy. In fact, the passing weeks made it clear that he was a very large boy. Marsha was very uncomfortable toward the end. Shortly before Graham's expected arrival, Marsha fired her original pediatrician and hired Dr. Halama.

Dr. Halama was as ugly as the toad he resembled. He was clearly gay and wore a lot of gold jewelry. An expansive man, we both liked him a lot and he became very significant in Graham's life. He was a practitioner of the LeBoyer birthing method in which the baby is cradled in a tub of body temperature water immediately after birth, the lights are dim, and sounds are muted all to make the new world more like inside the womb. All that worked to perfection at Graham's delivery. Marsha's water broke well before dawn July 2, 1980, and I drove her down the few blocks to our hospital with its third-floor birthing room.

It had open windows looking out over the Hinsdale train station, from which tens of thousands of commuters boarded for the ride into Chicago. Waiting with Marsha till she dilated fully, my attention kept returning to the trains whose slowing rumble, hissing brakes, and opening doors announced the start of a new day of work. It's hard to imagine that these commuters were productive in any natural sense. They did not herd, or grow gardens or gather sustenance. Some of them, tradesmen, built things, buildings for commerce perhaps. Other tradesmen operated the utilities or prepared and served food for the least productive of the commuters. The tradesmen mostly boarded the trains closer in to town than Hinsdale. Here the passengers all wore suits. Our interconnected world required their collective journey. Our income was completely linked to but far less than the stunning millions these non-producers made and spent. Each boarding was followed by the slow, smoky pulse of the great engines as they overcame inertia heading east with the dawning world.

With my newborn son, Graham, and Arena.

Marsha did dilate normally and in the predawn she was ready to push. Like Arena, Graham was sunny side up, but clearly much larger. Dr. Halama had Marsha breathe through her contractions to minimize their effect, and in between contractions with his hand on the hidden Graham's head, he patiently worked to flip him over. I have no idea how long or how many tries it took him to accomplish this, but at 7 a.m.,

Graham was delivered normally in proper alignment. He weighed nine pounds seven ounces. I was instructed to cut the cord, and then I took him and put him into the warm bath.

With utter placidity his widely opened eyes surveyed me, his first vista. He purposely moved in the water and in my hands. I held him there for a seemingly long time before giving him to Marsha. She had been pulled inside out. Dr. Halama called for other doctors and nurses to see that her cervix had come out along with Graham, and then said, "Let's see if I can put her together again." Fortunately, this was not a serious procedure, and Marsha and Graham left the hospital the next morning.

By the second day, Graham was following people about the room from his crib. We were proud of how vigorous and handsome he was. He sucked at his mother's teat like there was no tomorrow. His cheeks extended like a squirrel's, and we nicknamed him "the tanker."

Marsha's parents and her siblings were thrilled with the baby, and we spent a lot of time at their wonderful Hinsdale home. That is, we spent a lot of time there Graham's first month. Marsha had done something incredible. She wanted to return to New Mexico and asked me to find something. I called long distance to New Mexico Catholic churches and found a teaching position at the San Diego Mission School in Jemez Pueblo. The pay was $8,000 a year, half of a public school salary, but then I only had lots of credits, no certificate. That was enough. On our trip back to New Mexico, we drove the loaded van directly to the San Diego Mission School in Jemez Pueblo. After meeting the sister and taking a brief tour, we drove four miles north and turned into a secondary canyon and drove to the end of the pavement to Canyon Landing, a cluster of cabins, the last place on the road before tunnels into the Jemez District National Forest. Our two-bedroom, half-log cabin cost us $100 a month.

Marsha's mother and father were quite sad. They already knew that I was on no fast track to success. It also turned out to be a three-year retreat from my horticulture career although I continued to garden and learn about plants in nature.

Little Graham's first days in New Mexico were pretty idyllic. On warm days we walked down to the tiny Guadalupe river and he bathed and rolled and played in its clean current. He soon took to the backpack and tugged on my hair and ears like a bad rider tugs on his horse. After the first few months in bed with us, I built him a wood-slatted crib. It hung from the ceiling and had a commanding view of the sleeping parents below. He became the imperious ruler as soon as he could pull himself up and stand. His grandparents, aunts, and uncles all traveled out to see him. He lived in this idyllic setting until he was three, and in that time, both his younger sisters were born. I gladly tell the story of coming home from work and finding him inconsolable, crying to his mother, "I want my Da."

His idyllic life continued through Flagstaff and onto the Pacific beach in Mexico. He turned five shortly after we had ensconced ourselves in a nifty little home there. Three miles of wide sandy beach beckoned outside our picture window. Our elevation was eight feet. At high tide, we were 400 yards from the foam. At either end of the beach, rocky headlands were full of anemones and little octopi. All of us witnessed an octopus we knew being killed by someone who claimed they were going to eat it, sad. At night Graham listened to our readings and learned to read books on his own. There was no television to distract him from the joy of growing up in such a place.

He was seven, and halfway to eight, when he started first grade in Hitchita, Oklahoma. His teachers absolutely loved him and kept saying how much better it was to have children be older when they start school. Of course, Graham had been read to since he first could distinguish voices. Taught by his mother in Mexico, he already knew how to read, and physically, he was more than a match for the rough and tumble Oklahoma farm children in his classes. Graham's physical stature belied his birthweight. He was one of the smaller boys as he grew up, but a late spurt established him with above average height and a muscular build.

In Oklahoma, Graham and I drifted apart. He was a Cancer, and that is a difficult sign for me. He is much more harmonious with Marsha's Capricorn. When Marsha became the breadwinner, there was a sudden change in my family role. I was no longer the provider and Marsha began, as I sensed it, disrespecting and disregarding my role in the family. There was a national movement in those days tearing down the role of the patriarch and men in general. Archie Bunker and "Meathead" typified it. My commute added three hours daily to fulltime work, and when I was with Graham, it was usually on a trip when Marsha became her most authoritarian. In high school, Graham seemed embarrassed or uncomfortable when I cheered too loudly at his basketball games.

He only went to Midway High School two years, however, because after his sophomore year he was accepted at the Oklahoma School of Science and Mathematics. This was a state-run boarding school in the capital that admitted the 75 top Oklahoma high school juniors for two years of exceptional instruction. Graduates of OSSM enter college with twice the math and science of public high schools, all of which was taught to a much higher standard. The year Graham graduated, his school had the highest average ACT score in the nation. It was and remains that good. Graham earned a pretty much all expenses paid scholarship to Oklahoma State University and took his degree in mathematics. He then went to Chicago to earn his master's degree in Applied Mathematics at the University of

Illinois, Chicago, and to meet Katrina. He had a couple of years to get to know her better as he worked with a trader/investor at the Chicago Board of Trade. He and Katrina then moved to Bryan, Texas, where Graham picked up his PhD, and Katrina advanced her education in public medicine.

I got to see Graham and Katrina several times when they were living in Bryan. In 2011, Graham and Katrina had a spectacular wedding at a local winery there. Pictures from that evening showed me at my heaviest ever bodyweight. I had ballooned all the way up to 187 pounds. Probably from lots of beer drinking. At the wedding I drank a lot of wine, too much. I owe Graham and Katrina an apology. Here tendered.

From Aggieland, Graham and Katrina moved to D.C., where Graham works at a think tank and Katrina with the National Institute of Health. In April, 2017 Katrina delivered our first grandson by blood. And they named him William Nash, forming a link with his father, William Graham, and his grandfather, William Sand. Just last month Katrina and Graham added Nora to their family. It is heartwarming to see that my son's family is thriving.

Dracaena warneckii. The dracaenas we owned had to be kept in the house in winter. They are native to the bottom tier of rainforests and survive in little light. In a pot they can go months between waterings. This is why: the tropical house plants, dracaenas, philodendrons, and palms grow in the forest floor in deep shade. This is why they can tolerate the low light interiors of our dwellings. Their roots are very superficial and remain in the duff never entering the soil beneath where the giant trees suck up the 200 inches of rainfall. It is moist warm and dark in their environment. They transpire very slowly to match the light. In a pot with heavy potting soil, they can go seemingly forever without being watered. Succulents, which grow in 15 inches of rainfall ,I water nearly every sunny day latterly 20 times more often than the dracaenas.

16

Jemez Pueblo

I got the job by calling long distance to the New Mexico Catholic diocese and inquiring about teaching openings they might have in their mission schools. There were two openings, and I knew both locations. Penasco was high in the mountains north of Espanola and halfway to Taos. The other, Jemez Pueblo, was more appealing. It was one of the larger Pueblo tribes in Northern New Mexico. There were about 2,500 Jemez people, and they spoke their own unique language, Tewa, as well as being fluent in English. Jemez Pueblo lies on the Jemez River after it has been joined by the Guadalupe River, waters which flow from the southwest flank of the Jemez Mountains far from Espanola. Some of these waters are infused with the heat and minerals of the long-dead volcano. The pueblo is in chaparral land, almost desert just outside the mountain. It lies at about 5,500 feet elevation, and from there, the Jemez River flows about 50 miles east before it enters the Rio Grande near Albuquerque, 500 feet lower.

Their reservation was substantial and well-watered enough for irrigated farming. They had lots of access to the contiguous national forest for hunting and grazing. In this way they provided much of their own sustenance. There was no oil or gas in Jemez, as in Navajo land, so the tribe itself worked hard for a modest level of prosperity. Women were known for their pottery and for the bread from their hornos. The hornos were hand-built adobe structures that looked like small igloos. The ladies filled them with hot and quick-burning cedar. As soon as it flared out, they swept the ashes, put in their yeasty white flour loaves, and covered the opening with a blanket. Some of them set up shop under handmade arbors roofed in cedar branches along NM State Road 4.

Their dances and festivals, the public ones, brought in the Navajo from Gallup and the tourists from Albuquerque. There were also Jemez men who drove the 50 miles to Albuquerque for employment or found local work with the Forest Service. I am not sure if there were paid positions within the tribe; there may have been just one or two. Two or three Indian mothers worked as aids in our mission school, which was K-8 and had about 100 students. Entering this community as a school teacher brought me into a contact with the Jemez culture that I could have achieved in no other way. Like the Spanish, the Jemez have irrigation ditches and grow similar crops, I am sure, with their own special corn. After absorbing something from the Jemez' long-gone cousins in Bandelier, I would now meet all the children and all the parents of a living culture and be a figure in their lives.

Marsha, the two children, and I drove the 1,300 miles from Chicago to Jemez in late summer, 1980. I remember how pleased the principal, Sister Michelle, was to see me. She hoped a well-formed, normal-looking family man could manage classes of wild Indians. Previous nonreligious had not done well. I was to teach seventh- and eighth-grade math and science. My partner, Sister Mary Donna, alternated classes with me and taught English and Social Studies.

There was a remarkable arrival at our new home. Driving out we had no idea where we could rent. Yet I remembered or sensed something as we drove on north up toward Redondo Peak, turning unerringly into a side canyon about four miles north of the Pueblo. I didn't know at the time but the little creek at its bottom was the Rio Guadalupe. About six miles up canyon was a fenced encampment of about 20 acres and six or seven half log cabins. They had been built in the 1930's as a logging camp. We pulled in and rented a two bedroom for $100. It had bath and amenities with a wood stove for heat. We painted the concrete floors battleship gray and liked them just fine. It was perfect and beautiful. Our landlord was Henry, a part-time school teacher and holder of a 99-year lease for the property.

We were at 6,200 feet elevation and just one-half mile up canyon were the locally well-known Gilman tunnels. These tunnels allowed a roadway through an otherwise impassable box canyon. They had been originally built for a narrow gauge railway which was used to haul 200-foot-tall ponderosa pine trees down to the camp. When these trees were tapped, they were the best remaining pine nationwide and they fed all the construction east of the Rockies. Believe it or not the rest of the East had been logged out by 1930. They may have taken all the trees on the Jemez mountains in the years they operated, I have no idea. The ponderosas must have been replanted, or they reseeded naturally after the carnage. There would have been a great gout of erosion either way. When we arrived, the mountains had re-cloaked themselves in pine between 30 and 80 feet tall. Land wasn't bare and there were productive meadows and cattle leases. Much of it was very beautiful. Everything past the tunnels was in the Jemez Ranger District of the Santa Fe National Forest.

From our cabin, we walked 300 yards down to the Rio Guadalupe. It was set amidst red willows, not unpleasant, about ten feet tall. Occasional cottonwoods rose up along the course of the river within our view. The river was between six and 15 feet wide and was no more than a foot or two deep. If we hiked upstream, there were deeper pools, four feet, with little waterfalls you could hide underneath. We bathed in the river and thrilled to its ionic life. I was more of a naturalist, no longer a horticulturist; for the next two and a half years, I was a school teacher. Marsha brought in just a few part-time hours to supplement, and financially we somehow made do.

Actually, the first year was good for all of us. Marsha made friends with ladies in the Pueblo and with the smattering of Anglos, like ourselves, hiding from the scourge of normality. A little school was started at the camp. Arena became the model student. We hiked the mountains and mesas nearby. Arena kept up, and Graham rode in my backpack. We built campfires and slept in a tent. In the town of Jemez Springs, twelve miles up the main canyon, there was a Zen Center, hot mineral bathing, and a public hot spring of 105 degrees. That spring was the only crowded place around, and we only went once because of that.

However, some ten miles up the mountain at about 8,500 feet elevation was a perfect circular warm spring 50 feet across and three feet deep, 82 degrees all year, even with snow all around. We spent a lot of time there and once came back down with a dozen colorful guppies which thrived in our fish tank at home. Someone had stocked the pool with these tropical fish, and they had proliferated. Mother Nature fooled.

At the mission school, I did reasonably well, though I clearly lacked organizational and disciplinary skills so important for successful teachers. Marsha, the children, and I attended the feasts and public dances at the pueblo. My students danced wearing Indian regalia and made up with either blue or orange powder. They did not like being noticed by their teacher in this setting, so I avoided catching their eye as they did mine. Adults we knew in the tribe always invited us to "come eat" and we enjoyed a lot of delicious red chile dishes accompanied with bright jello in their kitchens. Feeding, even Navajo strangers, was a custom at Jemez at these festivals, and all the Navajo and many of the Anglos knew they could enter any home in the village and be fed. I was amazed to discover that at eighth-grade graduation the teacher was expected to come eat at every graduate's home.

This first year I ate at ten different homes, and the second I actually ate 24 small meals with 24 sets of parents some of whom I was meeting for the first and only time. At the end of the first school year, I was rehired for 1981-1982 and our summer was free. Then with great good fortune I was hired for eight weeks as director of our Forest Service district's Youth Conservation Corps. I actually only worked five weeks because we took a long summer trip back to Chicago, where we luxuriated in the love of Marsha's family and their swimming pool.

Upon return I took up the YCC job, and it was a very interesting one indeed. So much happened in just five weeks. The Youth Conservation Corps was a Forest Service grant program. I am not sure how many years it ran, but this happened to be the last one. Our district had about 20 high school students all from the Jemez Valley Public School. They were well divided between Indian, Spanish, and Anglo. Their 40-hour weeks were divided into three appealing categories: education, conservation, and recreation. The lessons at the ranger headquarters were very good. The young rangers and I gave lectures, augmented by the best of environmental and forestry films borrowed from the greater US Department of Agriculture library. The conservation was actual hard work in the mountains; a lot of it involved building small anti-erosion structures or working with water retention in some way. We also constructed pathways for hikers. We had a couple of vans for transportation.

The recreational part of the program took us to a walking path to Las Conchas, the Shells, from the State Road 4 marker. It lay at about 9,000 feet elevation, half a mile below Redondo Peak at 11,254 feet, the tallest in this half of the Jemez range. A strong mountain stream had cut through the pumice leaving several large deep pools shaped like shells. Above one deep pool, a brave person could jump or dive in from 15, 25, or 35 feet. As a kid I had been afraid of high boards, now nothing could be more exhilarating for the 35-year-old kid leader of high school students. It sure was fun. I imagine the authorities of today would have a heart attack if they saw how much the children enjoyed their semi-dangerous fun. We were all up and over the mountains, then down into the cool shade and water; to learn a mountain, that is a big thing.

My mountain summer ended on a curious note that has stayed with me ever since. I had become friends with the three or four young professional foresters who manned the bottom echelon of our district. Their education had been about proper management of forests. The dangers of over-cutting and erosion were well known to them. All declared to me that the Jemez had not yet recovered enough from the harvest of the 1930's to support any logging. This was their position when the harvest orders came down from above. The country needed lumber; Jemez must help.

In addition to the young rangers, I also knew the district manager, who was a conservative career forester in his mid-forties. When word came down about the cut, my young friends were vocal in opposition, and I saw the manager's eyes go completely cold and lifeless. In that flash, I understood how all government worked. In a chain that extends far, out-of-sight world rulers issue commands and in order to enforce them, their underlings, their sycophants, had to let their eyes go dead. What any employee below them might say in contradiction of the command, even if it had the force of law or policy, would not defer the command. And if they were bothered enough or frightened enough of what you were saying or knew, they well might kill you. In this empty glare of the district supervisor lay the coverup of the JFK murder, the Vietnam War, the Cold War, the mining, the clear-cutting of trees.

From my first full year of teaching in Jemez Pueblo, I internalized a fact which came to my attention via two of my seventh-graders. One was a beautiful, smart, and athletic girl, Yolanda Toya. My relationship with her was completely proper, but she could certainly qualify as the teacher's pet. As a teacher I initially liked all my students, and my pueblo children were well formed, appealing young people. In that setting my top girl turned in one outstanding paper after another and was bright and participative in class. Suddenly, with the arrival of some major grading period, she forgot everything that I was sure she knew, and her scores fell dramatically towards the class bottom. I perceived that she could not allow herself to be the teacher's pet or rather to excel in a path that separated her from friends and clan.

There was another student, a boy, Ivan Gachupin, who initially I took as a slow learner. He wasn't very interested in class and his papers showed it. Then out of the blue on some test or other, he turned in a really good paper. Maybe he cheated, but it seemed to come from him, and I felt a direct communication from him saying, "I am a smart person, and am letting you know that. I do not take your assignments seriously."

I quickly fit these two examples into the bigger picture of Jemez, which made it impossible for a tribal member to get a PhD, an MBA, or an MD and return to work in the pueblo. Everyone knew that to take this path was to leave family, clan, and tribe. That took care of the motivation question. I did my best to teach only for a love of knowledge and for the students to find satisfaction and self-worth in the solution to a math problem.

In my second year at the mission school, I had a big class of 25 eighth-graders. Among them was Joseph, who was the uncle of Norman. These two were the dominant male class members, the leaders of the pack. Norman was young bull of a kid; Joseph was slender and kind of stood behind Norman's leadership. Their resistance to my "good classroom order" was too much for me, and I suspended them and five other boys, their followers. The suspension required the parent or guardian to come in for a conference about their child's behavior. Sister Michelle appeared to back me, but kept me out of the conferences.

Marsha and I, and the Van

In retrospect I find that quite unfortunate. It turned out that the pueblo had the contractual arrangement with the school that education be provided for Jemez children period, that they were not to be suspended or expelled. It also turned out that the guardian of both Joseph and Norman was an important figure in the village. He rebuked sister, the school, and myself by removing the two from the mission and sending them to the public school, which they would attend for high school the next year anyway. I now know that the San Diego Mission did not continue for too many more years after we left New Mexico forever.

Later that year my class' misbehavior was so disruptive to my concept of teaching that I refused to be their basketball coach. Poor Sister Mary Dona had to end up doing the job, and I bet she was better than I. By adding to Sister Michelle's hardships, I pretty much sealed my teaching fate there, and I knew that there would be no third year for me at the mission.

In October of 1981, my two dearest friends from Teacher Corps, Vince and John, arranged for a weekend meeting at Mesa Verde National Park in Colorado. Both of them were close to Marsha as well; both had been in our wedding, Vince the best man. The weather was beautiful sunny fall; being with my friends was bittersweet, the ruins, somehow disappointing after Bandelier. Vince had a fatal rare kind of cancer, and I never saw him again, nor did we talk much after that trip. He died about a year later. The real event of the trip was making love with Marsha in the van and creating Heath.

Here's a view down through the glass roof of the greenhouse.

Dill always took off in the greenhouse from a fall seeding. I never had much luck with it outside. I also always had rosemary and thyme growing in the greenhouse.

17

End of the New Mexico Dream

My teaching load at the high school was enormous. I had five classes totaling about 120 students with four different preps. I remember having to work very hard each day and being tired and relieved after I got home. I had this notion that students could learn to really enjoy math if they were successful at it. Each class would begin with a short lesson, followed by an assignment, followed by my moving the rest of the period from desk to desk helping students individually. I know that they liked that, and that many of them learned. Unfortunately, by moving and working in that manner, I left myself in a weak position for enforcing discipline. Some classes went well, others not.

Other teachers did not work as I did at all. One respected old-timer had his classes copy from board or book the entire period each day. He maintained a reputation of one who would paddle hard if necessary. I watched another teacher through a little window; he reclined on his chair with his feet on the desk facing his students. They were seated as far from him as they could get. His assignment likely was long and in no way challenging. All the teacher had to do was monitor behavior and have punishment ready. At the end of that same day, I was in the office when he came in, yawned, and said how bored he was. Any assignment the children did not already know how to do was asking for trouble because the students would have questions or confusion. That was the way the whole school was run and except for sports, faculty trysts, and teen love; nobody wanted to be there.

The most difficult of my five classes arrived after lunch. The thirty-one students taking ninth-grade general math included Norman, Joseph, and a 200-pound sociopath child whose name I no longer remember. Norman, now a ninth-grader, was considered the toughest kid in the school after decking an eleventh-grade thug with a single punch.

It took till December for the explosion to arrive. I began the class with a hardworking lesson. I believed the class was paying attention, when Norman burst out, "No one's listening to you, Mueller." I recall pointing my finger at him and saying that his parents would be coming in. He jumped up from his seat and came right into my face saying, "The hell they will." Just that fast I slapped him on the cheek hard, and then I backed him twenty feet into his desk, cursing him the whole time and daring him to hit me. You can imagine what that did for my teaching career. I wasn't fired right away, but the next six weeks were awful. Now every tough boy in the school tried to goad me into a confrontation. My fifth period got worse, not better; Norman more or less left me alone, but the sociopath got worse and worse, getting out of his seat and threatening weaker students. Eventually, I made another mistake giving a light whop to the butt of a student who was crawling around between rows. I was fired the next day.

This, of course, was a massive blow to the well-being of my family. There were five of us now, and Marsha was pregnant with Rachael. We had no income. Although my career as a teacher was in shreds, I still had the option of taking classes toward a math certification. By doing that, I was able to get a student loan for $4,500 which allowed me to buy a 1965 Valiant, which I used to commute to Albuquerque, where I soon found a job with the city's best known and oldest florist, Nesbitt's Flowers. I also took three difficult math and computer programming classes toward the phantom teaching certificate. I say phantom because after a while, I realized that with the firing on my record, I had virtually no chance of being hired by any other district. I remember doing well in my calculus classes, the ones I should have taken in 1965 at Rice. I now have a kind of love for math and the certain knowledge that I could have had a very good-paying career with that in my pocket.

Shortly after the teaching disaster, our landlord, Henry, decided he wanted to sell off the cabins for cash and to end the small community of which we were so much a part. First, he sold the school cabin and then forced out the teacher, Sara Roemer. Then he came after us with draconian new rules designed to make us want to leave. One evening he came around with his six-foot six-inch friend and was so rude that I informed him in a voice that left him trembling, that if he did not leave immediately, I was going to knock every tooth he had right down his throat.

It was a strange sensation; my almost skinny 140-pound body was coursing with such an anger that had I knew I would destroy this man should I release it. Marsha told me Henry began trembling as I threatened him. He certainly did not make the gesture of resistance that would have released my fury. Both men stepped back and quickly left without a retort. I didn't know I had that in me, but afterwards I ceased to have the old doubts about my "bravery" or lack of it. It also required us to move again, and we rented a home in nearby Ponderosa, where Marsha and the children had a tolerable existence the last few months that we lived in the Land of Enchantment.

That time was defined by the nine months I worked for Nancy Nesbitt. Nancy, in addition to her florist shop, had a commercial plant maintenance business that required one full-time person and two sizable greenhouses. One house served for tropical plants trucked in from Florida, the other was available for production. She put me in charge of her flagship account at the bank and the greenhouses. Both employs were interesting for their aspect of horticulture and in the case of plant maintenance, human interactions. I grew a fine spring crop of organic vegetable starts, and about 200 beautiful 12-inch hanging petunia baskets that all sold for the astonishing price in 1983 of $17.50. Too bad for me that Nancy's business was slowly sliding under. She had some illness, bless her heart, probably cancer. She paid me a very low salary, but for the first seven months, her need for me was so great that I earned enough overtime to maintain our family existence. Adding in the long commute, I was rarely home it seemed, and I had a kind of second life in Albuquerque, which included flirtations that brought my sexual desires into play, this time not leading to actualization.

When summer came, Nancy could offer me no more overtime. At that moment, my father and my brother-in-law offered the job that once again suspended my horticulture career and took us to a farmhouse in Doney Park, just outside Flagstaff, Arizona.

Do you like Siberian Irises? I grew a few dozen one-gallon pots of them and sold them all. Of course, it took three years before they bloomed.

What a great collection of succulents with a hanging spider plant and the massive star jasmine.

Pallets with my outdoor nursery crops were scattered around house and greenhouse. Those are the grapes in the background.

18

Heath

During Marsha's pregnancy with Heath, I finished my second year at the mission. We continued to love the enchantment of living in this beautiful setting and being by the river with family and friends. At the end of the school year, with Heath due in June, Father Emeric, with an out-of-the-blue generosity gave me $2,000 as a kind of severance pay from my $8,000 job. At that same moment, I was hired by the Jemez Valley Public School as a full-time math teacher. The pay was for $16,000 a year. It seemed like a fortune, and for the first time ever, I was earning a salary I need not be ashamed of. It came in that wonderful place where salaries like this were not plentiful. It was as if every prayer had been answered.

Heath was born on June 12, 1982, in Albuquerque. Marsha had a lady doctor who was comfortable with our new age beliefs; she had not done the LeBoyer warm water birth before, but helped us arrange that. It was an hour to drive in after Marsha's water broke. There was plenty of time, and she dilated normally not long after we arrived. After that nothing happened. Marsha apparently had no musculature left to push Heath out.

After a "too good a while," the doctor ordered pitocsin, a powerful birth stimulant; then somehow after about eight hours of labor, Heath was delivered. I cut her cord and placed her in the warm, not quite warm enough, water. But at that moment, I had no attention left for Heath or water or anything, but the sight of Marsha bleeding. I could see a flow following each of her heartbeats and her blood pooled on the floor at my feet. The nurse with trepidation in her voice called out the blood pressure, 50 over 32. This was a horrible sight for me. It overwhelmed my senses, and it shocked everything I thought I was; it wadded the writing of my life story and balled it into the trash.

The doctor stopped the hemorrhage with her hand, somehow bandaged or sutured it, and Marsha passed out of the danger zone. She would have died had she not been in a hospital. She was given two or three units of blood that kept us worried about HIV for years after that. She and Heath were released the next day.

Somehow, Heath became a normal being after the terror of her birth. She nursed for about six months, the least among her siblings, but she must have gotten all she needed because she was a healthy eight-pounder at birth and from a young age, mobile and active. Shortly after she began crawling, she slipped away. It was a sunny spring day. The five of us were home, and Heath disappeared. I don't remember who found her; she had traveled through gravel and sage over one hundred yards and seemed to enjoy doing it. Finding independence as she did was a great accomplishment that allowed for many things later on.

Otherwise, she was a distant second princess to Arena, who was the first grandchild of both extended families and in the shadow of the first son, Graham, the Prince whose accomplishments were regarded as awesome by all his family admirers. Heath never seemed to notice. Maybe it was because Heath was a Gemini; she was lightweight, supple, evanescent. I thought she was fragile, balanced on an edge. Being Gemini, she and I were astrologically well connected, and in form, she was by far the most similar to myself; the others were "Heasleys."

When Heath was four years old, we were living on a Pacific beach with three miles of sand between rocky headlands. The six of us had the beach to ourselves one sunny winter day, and I decided to stage a race.

I already well knew the movement of Graham, the strong and sturdy, and Heath, the nymph. I drew lines in the sand for a race of about 200 yards. Graham ran like a young soldier of fortune or marine. I could almost imagine him in combat uniform carrying his weapon. He lost badly, six years losing to four. Heath's feet hardly seemed to touch the ground, her entire body seemingly in total relaxation virtually without weight in a kind of physical perfection of essence. It was a pure primordial sense of body that came before the weight of the world oppressed it. Her running reminded me of Greg Peters' race twenty-one years earlier.

I guess I am saying that something essential is lost as personality stifles essence. The fluid unhindered movement of her four-year-old body was analogous to the four-year-old emotion I described earlier about myself in Dallas. Ten years after we left Mexico, we were living in Oklahoma, where the state had established the Oklahoma School of Science and Mathematics for the state's brightest high school juniors and seniors. Tuition and room and board were free, but the course of study was rigorous beyond even the best private schools. In 1997, Graham became one of their early students, his admission fueled by a very high ACT score. After he completed his first year, Heath, on her own, made application as a "too young" freshman. She wasn't accepted that year, but easily made it in the next, following her brother's graduation. It turned out that her determination was stronger than seemed possible from my misgivings about fragility or the inheritance of her birth.

We were sitting together, talking about this chapter at her home recently, when she told me that growing up she had gone first to me to share her stories, to ask her questions, and to get support and advice for her difficulties. She shared a story of which I had no memory. I had come home from Tulsa exhausted from long hot day of work and drive and had gone to bed early. She was being bullied in school and came in as I was nodding off. I am sure she did not want to talk about it with her siblings around. Apparently, I sat up, woke up, and we talked about what was going on. It seems like most fathers would do this for their children; most fathers would want their children to talk to them especially about being sad or scared. So it didn't seem like I did anything special for Heath. But it must have been special to her because as she told it, she soon had both of us crying. I am sure that this moment between us is in the penthouse suite in the hotel of memory. It's really not so hard to win the love of your children. It gives one sorrow to consider families without fathers.

Heath rode her determination and her ethereal nature on beyond the great achievement of OSSM and is now a successful child psychiatrist and mother. Her on the edge fragility has been totally overcome, and now in my waning days, Doctor Mueller-Sears has lifted "my" fragility and helped prop me up so that I could tell this story.

This is Heath's daughter, Wednesday Nicole Sears.

19

Doney Park

My sister had married Dixon White, a circus performer, a ringmaster, a Maître d', a martinet, a striking young man with a devil's beard. He was the center of attention, a mama's little darling with intensity. He acquired me almost as a prize. He rescued me from disaster and took over my intelligence for his own purposes.

After their circus run, Dixon and Annie had settled down in Flagstaff and opened a Rainbow Vacuum Cleaner business. Rainbow is actually a very fine product, but the company operates pyramid-style. To make a living as a salesman is rare, though my sister became a notable exception. In the fall of 1983, they had just completed their first year in business, and they felt a need to hire a Jack-of-all-trades office manager, repairman, phone answerer, sales trainer, and occasional salesman. This person would be me, and I would be paid a salary of $20,000 per annum. I was to rent an old farmhouse, which was on 10 acres of property they owned, along with their new home which included their Rainbow office. Oh, and the farmhouse was bordered by national forest. My mother and dad were great admirers of my brother-in-law and let me know they would be very disappointed if I didn't snap up this opportunity. So, of course, that is what happened.

Doney Park was just outside of Flagstaff, Arizona. It had a spectacular but distant view of the San Francisco Peaks, whose jagged trio looked clearly younger than Jemez or even Truchas. Throughout the winter, they loomed as frozen bastions, and howling west winds blew a curtain of snow over them and down toward my gazing face.

Doney Park itself was bleak. It was a former dry bean field, and the recently installed, well-heeled homes had not yet landscaped themselves into anything appealing. The adjacent forestland was also nothing special. It was a uniform stand of young, perhaps 40-foot, ponderosa pine. The ground around and under the pine was not well covered in vegetation; I suppose it had been clear-cut then replanted. The ground had a lot of scoria in it, like from a more recent volcano. So much so, that it fostered a forest anomaly.

Everywhere in New Mexico and Colorado, pinyon pine grow at lower elevations than ponderosa pine. The ponderosa, known to use and need more water than the pinyons, thrived on the 25 inches to 30 inches of precipitation normal for 8,000 to 9,500 feet of elevation. The pinyon pine typically got 20 inches to 25 inches at their range of 6,000 to 8,000 feet. Rising above the ponderosa pine near me in Flagstaff were mountain hills or uplifts, and those places were covered with pinyon pine, everywhere upside down from the rest of the Western mountains. Apparently the gravelly lava soil in that part of Arizona drained so rapidly that only the drought tolerant pinyon could exist on the uplands. This seemed significant to me. I often visualized the flow of water around me, above and below ground and from the air as well.

There were some really beautiful places near Flagstaff, particularly Walnut Canyon and Oak Creek Canyon, whose springs gathered from underneath the scoria. We enjoyed those and several other offerings of nature there, but left too soon to really explore or "know." We never even went to the Grand Canyon.

As in New Mexico there were many intriguing remnants from the former inhabitants of the land. The farm house though was unappealing and drafty. Doney Park in winter was just flat cold and windy. Marsha did not feel it was safe to use the unvented propane stove at night.

Mornings the temperature inside the house was in the thirties, and I remember our record low of 34 degrees inside the house. Rachael was six months old, Heath was twenty months, Graham was three, Arena was nine. Marsha took Arena to Montessori school, and when money got tight, my bosses complained about the expense.

None of us liked living in Flagstaff, nor was work anything beyond ordinary for me. It was not enjoyable like all the many different horticulture jobs I held. When I first arrived, there were about a dozen salespeople of whom only a couple were productive. One was a Navajo fellow named Tex, who had sold over 100 machines on the reservation and up near Tuba City. Rainbows were expensive; we charged $725 cash for each and offered a couple of different credit plans. Part of the pathway to success for the office was to recruit as many salespeople as possible. All were trained to give a convincing "demo." The initial trio of sales made by these recruits was recompensed at a very low $50, and it took a while to achieve the $200 commissions that the real salesmen got.

My brother-in-law told me that the recruits would sell their parents, siblings, other relatives and best friends before they realized that being a Rainbow salesman was not a real option. It was a more a strategy for penetrating the upper-income strata than seriously recruiting employees.

My arrival coincided with a downturn in their sales. My analysis was that Annie and Dixon had basically sold out the entire Flagstaff market in just a couple of years and the cherry-picking was over. Just after the sales peak, Dixon bought Annie a red Mercedes-Benz convertible to go with the green XK-E he had driven through Chicago three years earlier.

Before too long our presence and my salary turned into something inconvenient and awkward for my sister and brother-in-law. So after nine months living in Doney, my career and our lives took another huge leap into the unknown.

I highly regard oak leaf hydrangea. It stays attractive all year long. The variegated shrub to its left is false Solomon's seal, which was a slow-growing and attractive perennial for over five years.

I had so many species in the little greenhouse. The hanging staghorn fern eventually doubled in size so I finally had to pull it apart and divide it. The recovery was slow, but eventually, the six parts I divided it into would have sold for $25 each had I stayed in business. Here also are the grey aeoniums and a kalanchoe.

20

Rachael

Rachael was conceived at the high-water mark of my teaching career, when the possibility of staying in this beautiful enchanted place was in my hands. The possibility seemed majestic. Nature and history were so, so close.

I climbed the mesa just north of Virgin Mesa. They were the same height, but mine was smaller. At the foot of mine were partially standing walls, flint chips, and pottery shards. I never took any of that stuff. The ascent of the big mesa began with spying the vertical rift that had been formed by water pouring off the mesa. Here the going was steep and tiring but simple; there were always rocks for hand and foot holds and one little ten-foot cascade after another of dry waterfall with lovely level gravel to rest on. Just a few feet to my left and right, the mesa presented vertical cliffs to the world. The top of the mesa was at 7,800 feet, and the talus base, 1,000 feet lower. About two-thirds of the 1,000 feet was straight down.

Once on top, everything looked pretty similar, unbroken junipers, maybe a few pinyons; I've forgotten. It was neither interesting nor pleasant up there. I hiked to the north end of the mesa top, which I'm guessing was about 40 acres. From there, idly speculating, I realized that from on top you could not see a safe descent. Even a rill that seemed promising might end before the bottom, and 200 feet would not be better than 700. Knowing I had to find the point at which I had ascended, I scurried back that way and chose a route that seemed similar. Only it wasn't. The ground got very steep, the rocks were a little loose, and suddenly, I was eight feet away from that 700-foot drop. My legs shook as I crawled up and away from that precipice. No, Thank You. I went a little further and somehow saw the way down. Being stupid and not being very aware of my surroundings could have been costly up on the mesas. Most outdoor things we did as a family were safer than solo climbing, and Marsha and I were getting along well and loving our children.

The concern was that by November, when Rachael was conceived, I was already struggling with my teaching load, and when Marsha was two months pregnant in February, the worst came to pass and I was fired. Then Henry kicked us out, and we luckily found a very nice home we could just afford in nearby Ponderosa. I got a huge surge in money by taking a $4,500 student loan from the University of New Mexico, where I did indeed take three tough math classes. Most of that money went for us to live on, and we also bought a 1965 Valiant from a friend for $500. The school allowed me to keep the health insurance and that paid for Rachael's delivery in the same ward as Heath fourteen months previously. Marsha's pediatrician, with a strong clear memory of Heath's bloody delivery, waited till Marsha dilated, gave her a saddle block and extracted our fourth child.

Rachael was the one people scratched their heads about. I think some people thought Rachael was one too many. Nobody thinks that now. I found work at Nesbitt's in Albuquerque and that kept us in New Mexico until Rachael turned one, when we moved to Flagstaff.

She was another eight-pound healthy child. Rachael nursed the longest of any of Marsha's children, two and a half years. Towards the end, Rachael would lift up Marsha's blouse, point and say, "Toe tey?" Growing up, she became a strong child, fearless and headstrong, but in a way that none of us found offensive at all. Her Leo sun was bright and hot. Heath our long-suffering Gemini was fourteen months older than Rachael, and that was barely enough to protect herself from her assertive little sister. Rachael always seemed to notice things before the others, and then, before you knew what happened, she was saying, "I'll take that, thank you."

Contemplative types like Heath were easy meat for Rachael. Guess who went to the front of lines. Guess who was the center of attention. All five of us loved the baby. Guess who, upon the day of our invitation to lunch with our Mexican landlords and benefactors, the Fiembres family, disappeared from our sight amidst the bamboo, then appeared again, now running around naked in and out of the bamboo. At 21 months, she was plenty quick and agile enough to avoid capture so that it was only later when she came out voluntarily and allowed Arena to dress her that she could attend as a proper young lady. Meanwhile Marsha and I were sitting with Trinidad's two children, who were about our age with their own children. They were quite bemused, not just at the incorrigible Rachael, but also at the discomfiture of the parents. It turned out that there was a kind of bonding around that moment even though we barely understood each other linguistically. We never saw much of them again after that, just a wave, I think.

Rachael was not yet two when we moved to Mexico and had turned four before we left. So she won't remember much from there in her mind, but her body formed its memories from those years and those memories were in the healthful sunshine of a Mexican beach. She also got read to a lot and formed a strong vocabulary and ability to communicate. She got subliminal influences and stimulation from her three older siblings. When we moved to Oklahoma, she excelled from the beginning of kindergarten in Hichita just like her siblings.

There was a poignant memory I have from when Rachael was in first grade. At that time, I was working in Tulsa and our home was 65 miles south in Wainwright. But the Wainwright school basketball teams were playing in Leonard, Oklahoma, not far from Tulsa, and on my way home, I arrived a little after the game started, just in time to see Rachael steal the ball and then dribble and score to make the game 12 to 0 in Wainwright's favor. I sat down by a teacher I knew who told me that Rachael had stolen the ball six straight times and scored all the points in the game! This, alas, was the summit of her basketball career because after that game the league raised the goals to the full ten feet, and short little Rachael simply could not throw the ball that high.

She and Heath shared a bedroom in Wainwright and for the most part got along well. Eventually, one day Rachael's hot sun burned Heath a little too much, and Heath, uniting her twins, uttered an angry blast which scared even the little lioness. Heath gained the well-earned nickname of "Hell Cat" and thereafter, increasing respect from her now maturing sister.

Rachael, more than the other four, had to slog through Oklahoma rural public schools for her education. With both her parents locked into the grind and commute, little was offered her in the way of help or stimulation education-wise. When she was a junior, rather than trying to compete with her three siblings and go all out for the college world, she became a sort of stable, caring Mother figure for all her drug-addicted schoolmates who were beginning to have children they couldn't take care of.

This caring and nurturing quality, more than any other single thing, characterizes Rachael. I am so fortunate during my last days to have her fiery caring beamed on me, her father.

After a desultory semester in college, Rachael dropped out of sight for a while, and then suddenly showed up with Rick Rymel, her rock and future husband. On July 18, 2006, Rachael delivered my first grandchild, Raelynn Nicole Rymel. That day was also my mother's eighty-fifth birthday. I wonder what that means? Sometime after Raelyn was born, Rachael began working at Staples, bottom rung. But after a while, she got noticed by management, groomed, and today manages a large and successful store in Shawnee, Oklahoma. At home she has two adopted brothers Raelynn's age, gathered into the family – agewise that is two thirteens and a fourteen. In addition, about seven other children consider her a second mom, and from early on, two more boys from Rick's first marriage claimed Rachael as a second mother. One of them is now married and has two children. My youngest child became a grandmother at age 36! And I just recently realized that this first child of Rick's son Casin, named Rebel Rymel, had made Marsha and I "Great."

Rachael recalls her youth:

> From my earliest memories as the youngest of four, my childhood felt carefree, filled with adventures with my siblings, especially the closest two in age, Graham and Heath. My sister Arena was nine when I was born, so I remember her as more of a mother figure, and I called her Nan or Nannie Nan. I remember adventures on the beach in Mexico, then moving on to adventures on the farm. My parents were both very loving, I remember my dad, Sand Mueller, as a teacher, teaching and working with us on almost all things especially gardening and math, which I didn't like at all unlike my siblings. My mother, Marsha Heasley, was a workhorse. I have lots of memories of lying in bed, missing her tremendously because she was at work. She is an amazing guitar player, and from the time I was little, I remember singing and dancing while she played our favorites, "Drunken Sailor" and "La Bamba."
>
> I have always considered my early childhood as very pleasant. I don't have very many memories from Mexico and zero from New Mexico. My favorite memories from Mexico were sitting on the rocks on the beach and taking sticks and sticking them into sea urchins and watching them close onto the sticks – also going whale watching on a boat with a dear family friend, and I have memories of playing on the beach and our hammock. Moving to the farm in

Council Hill is where all the memories flood in, so many good memories, and some bad, like the time my sister and I got into a poison ivy bush and grabbed all the berries and broke them in our hands. I think being exposed to so much poison ivy as a child strengthened my immune system to it, and now as an adult I only contract it through open wounds.

We had a mimosa tree in the back yard we loved to climb on and play in. My oldest sister, Arena, also used to leave us notes inside trees, and we'd find them and think they were from fairies or goblins, which my sister thoroughly convinced me she was a goblin. Dad built a greenhouse and we used to play in there, making mud ice cream and mud pies with the mulch. Dad also had a very nice selection of whipping willows sticks to keep us small children in line. I once caught one on the foot for not paying attention during a math lesson. We'd play all day on the farm. Once, we got into the back of a truck filled with corn and basically went swimming in corn kernels. My sis Heath and I once decided that we were going to run away; this was after we colored our faces with permanent marker. So we ran into the fields. We lived on 88 acres there. After a search for us, our brother lured us back as I recall by acting like a puppy and we followed the barking.

Arena and Rachael

Council Hill has by far my most fond memories of my childhood. My father used to pick us up from school using chicken feet as his hands! As we had lots of chickens in Council Hill, which those said chickens were evil. I once sat in the car for 30 minutes after being chased by a particularly aggressive chicken. I remember crying and just wanting it to go away.

My father could be quite a jokester. Some of my favorite memories of my father are walking on his back, the doodie dot (small hand creature that would get us) and the pancake roller. Dad would make a bum bum bum bum bum noise while rolling on the floor or bed and would proceed to roll us into pancakes. After kindergarten we moved to Wainwright; by then as I recall, both parents worked full-time jobs and there was more time just us kids and my older sister who eventually had a  house right behind us. It was nice growing up in that town for me as we had lots of neighborhood friends to play with. As we began to grow, we encountered more chores and duties around the house, which I hated and often pawned off onto my sister who is notoriously clean and tidy. We shared a room so she often just got mad and cleaned it up herself.

My favorite memories are always Christmas memories there. One Christmas lying in bed, we actually heard reindeer hooves and jingles on the roof. Pretty sure that was my father. One year, he was in charge of wrapping all the presents, and it turned into a fiasco because of the poor wrapping and un-named presents causing confusion. Mom made Dad apologize to us all after it; Mom always made Dad come apologize to us after they had a fight.

Growing up my favorite thing that I waited all year for was our summer breaks going to Hinsdale, Illinois, to visit my mother's family. Nothing like spending the summer swimming, playing with cousins, going to parades and shopping with Grandma and Grandpa.

As I grew into an angst-filled teenager, I often spent hours laying in the clovers in our yard listening to the Cranberries and finding four-leaf clovers. My sister and I often did this, and our record was a six- or seven-leaf clover once and lots of fours and fives.

It was in Wainwright that we discovered Arena's Ouija board and spent an entire day using it and ended up freaking ourselves out so bad that my sister and decided to watch *The Little Mermaid* to calm ourselves down. We heard a knock at the door that proceeded to hard knocking or kicking; we looked through the peephole and it was pitch black. Convinced it was demon, we ran through the house screaming bloody murder, and trying to reach Arena on the intercom to her house. Just as we decided it was time to call 911, we heard a familiar voice saying it's just me. It was Arena pranking us, not knowing that we were already on edge due to our shenanigans with the Ouija board.

Growing up in the Mueller house, we always had animals, dogs and cats. We had some very beloved pets. Our family dogs: Wolfie, Spot, Meatloaf (Meatball we called him as a puppy) and Gibby (or Skipper as Dad called her) Paws, Elfie and Striper Sue (Our dear cats). We all grew up with love and learned to love all people and animals.

My father loved to play Frisbee with us, and the dogs. He was quite the Frisbee player. We often would load up and go to Lake Eufaula – we called the beach Wai Kee Kee – and we'd play Frisbee and swim.

Also I remember one visit, we picked persimmons and they were so delicious. We also grew up with the love of camping, and many wonderful camping trips to Fountainhead. I remember one eventful trip, we were rained out and spent the night in the station wagon, and Meatloaf decided to poop. It was quite the dramatic event.

We loved nature hikes, spent a lot of time outdoors. Both my parents were great cooks; my father makes the best scrambled eggs! His specialty dinner growing up, which wasn't a huge hit, was brown rice. I would say that all of the Mueller children grew up with a great appreciation of food and are definitely "foodies." Most of my siblings are also avid gardeners and have a great connection with my father. My

mother is also a great gardener, although her specialty is more floral while my father does it all.

I myself have a black thumb and struggle to keep even the most basic plants alive, although I have this year started growing lavender, and I have a couple cactus plants that I haven't killed yet.

As a teenager, I was definitely the problem child and gave my parents a lot of sleepless nights, especially mom. I am not sure why I choose to do some of the things I did, but through it all, they both supported me and even ended putting me into rehab and sending me to live with father in Houston, which was quite the adventure for us both. But I ended up returning home and struggling a little bit longer until I met my future husband and had my daughter the first grandchild in July 2006, which basically ends my story as a Mueller, as I transitioned into a mother myself and had a family of my own, but that is a story for another day.

I can say that although our family isn't perfect and we had our fair share of issues, there isn't another family I'd choose to grow up with.

This is an amaranth from Jamaica, which came to Tahlequah from seed smuggled in by a friend of mine. They use it in the famous dish, "callaloo." This one is about eight feet tall. It remained edible after it got big.

The healthful, but mysterious milk thistle. It kept appearing and disappearing, jumping around, gone one year, then spectacular the next. Collecting those little seeds for somebody's liver is work with leather gloves and pliers.

Stinging Nettle, I bought a 4" plant. Three years later it was six feet across. Each summer I began shearing it as the flowering racemes extended. I used as much as I was able dried for tea and soaked in my water drum for fertilizer. Like so much of what I grew it was hard to keep within bounds. Sinewy roots crept out from the center trying to claim an even greater patch. It takes a village to use all the medicine in this plant.

Tight little version of the old-fashioned bridal wreath spirea, reliable, striking, and nostalgic in spring, ugly in winter.

21

Mexico

Marsha brought home a copy of *Florist's Review*. Actually, that was the only trade publication I knew of then that had nationwide help wanted for horticulture. So the one I responded to said, "Wanted, Grower of cut flowers for nursery in Mexico. Can live in US or Mexico." The telephone area code was in St. Louis, Missouri, and the man I talked to was indeed the 49-percent owner of a cut flower range on Federal Highway 1D at km 58 in Baja California Norte. He hired me after our second phone call, offering to pay me $20,000 per year and letting me know when he would next be in Mexico for us to meet.

We were very hopeful and certain that we could live in Mexico and do very well indeed on that salary. The call came in April, and we left belongings and the 1965 Valiant with my sister. The parting was civil, but not comfortable.

Our van was the same one we had bought for Plants Alive in 1973, and it had lots of miles on it, but it was still strong enough to carry the six of us and our little ten-foot travel trailer. We left Arizona early in April of 1985 with $700. Driving into Mexico, it was clear that one would not live in the US and hold this job. We were nearly an hour on clear roads to get to the nursery from the border. There was no delay getting into Mexico, but for a commuter, the return crossing into the states often took an hour.

We drove a few kilometers past the nursery, which we saw from the tollway, and came to a three-mile sandy beach between two rocky headlands where a fee of $2 per night was charged to park or camp on the beach. We pulled in there, slept in the camper, and drove back to the nursery for the morning meeting. We found that the nursery was not open to the public, but eventually, we were allowed in. Around lunchtime, the American owner came out to chat with his new employee, or so I thought. I do not now remember anything about this man, only that he told me he was sorry, but that he had been unable to come to an agreement with the Mexican owner, and that therefore he was pulling out and had no job to offer. Wow.

I was aware that the Mexican owner, Jorge Salceda Vargas was there, but I was unable to meet him. I am not sure what we did the rest of that day, but we did come to a decision that I would return to the nursery later in hopes that the Mexicans would hire me.

The nursery was called Alisitos, which may have been the name of the creek that flowed by the nursery and/or the coastal canyon out of which it flowed. I was never sure what the word meant in English. So I did return that evening and spoke with Fernando, a 28-year-old playboy, who managed the Rancho for Jorge. We went to a nearby hotel, and I bought him a few drinks and had some myself. My Spanish was very weak, maybe a bit better than his English. Fernando was interested in Americans and hired me for $35 per week, the highest he was authorized to pay. Most all the other employees fell under the state category of agricultural worker and made about $20 per week. When I began work the following Monday, we were still staying on the beach, but were down to less than $400. At that point, I do not know how we could have given up and gone back to California for work. How would one do that with no job and four children?

A week went by. I remember enjoying the immensity of the carnations, the mechanics of the nursery, and the spectacular canyon setting. I had many talks with Fernando, mostly on whether we could afford to stay or not. After that week we decided that in order for us to remain in Mexico I had to earn $100/week, and while I continued working the second week, my focus, with Fernando's help, was to get an interview with Jorge.

Fernando, of course, had his phone number. All I knew was that Jorge lived in Tijuana, the famous border city about 45 miles north of Alisitos. At the end of the day on Wednesday, I waited in the office of the Rancho while Fernando took a shower in the office bathroom. I was surprised that Fernando would need to shower before making a call and when he got out, he pointed to the yellow telephone on his desk. He laughed and picked up the phone so that I could see it was without any connection, a dummy.

We drove to Rosarito Beach, twenty miles north to find phone service. Surely Primo Tapia, ten miles closer had a public phone, but I don't know. I think Fernando hoped to meet an American babe in Rosarito; he had a lot of cologne on for just me. While there we ate a lot of tacos. I was good with eating taco do lengua (tongue) but Fernando ate a taco de ojo; the eye was literally staring right out of the tortillas. As for any women, I don't think he was successful, but our call to Jorge was; he invited us to his home the following Sunday morning.

Armed with directions and nothing more, we arrived at Jorge's door on that fateful April morning. I recall that the "Candidato" lived in a decent neighborhood and had a nice but not extravagant home. I remember better that there were two men in the front yard who were astonished to see us pull up. Who knows what they made of my Spanish gibberish, but they had us wait while they informed Jorge of our arrival. Before too long, we were escorted to the anteroom where Jorge, his wife and two or three children of theirs greeted us and welcomed us inside for a lovely Mexican breakfast of ham, eggs, and fresh orange juice. Marsha thinks that they had totally forgotten about our arrival, something was said, a bustle, cooking aromas. The Salceda Vargas family were warm hosts. All four of our children back then had silky blonde hair, and our Mexican hosts could not resist feeling it or combing it with their fingers.

I had some photos of my work in the states, and Jorge indicated he wanted me to grow some of those flowers; even better, he agreed to pay me $100 a week, although I was to be paid equivalently in pesos. That turned out to be not so good for us later. Before we left their house, he gave me my first week's pay, 28,000 pesos.

I was the ingeniero of the Rancho and would work a 48-hour week like all the other employees. No one was supposed to know how much I was paid. We returned to our beach and rented a cute little house about eight feet above sea level, and from the tiny three-mile sandy beach, we faced the mighty Pacific Ocean.

Our house was one of a cluster of little trailers and cottages that were leased, as we found out later, by Americans who came down on weekends to the ocean splendor we now enjoyed full time. Suddenly we were Mexicans. It was thrilling, not to forget that I was a "Mojado, al reversa," a wetback in reverse. We remained in Mexico at the will of my employer, Jorge Salceda Vargas.

We found that Mexico offered almost nothing in the way of postal service. We received one letter from the general store in Ejido, Primo Tapia. The terminology Ejido is a very interesting aspect of how Mexicans have organized their society which is worth investigation by the curious mind.

We needed mail, and mail with stamps that our families and friends could purchase, so that one of the first things we did was to cross the border and get a P.O. Box in Chula Vista. Sweet View was a lovely city, a suburb, really of San Diego.

Chula Vista became a big part of our lives in Mexico, not so much for the mail but primarily for the library which allowed us to take out 100 books at a time. These books were a source of great family unity, enjoyment for the parents, and education for the children. Within a month or two of settling in, we drove back to Arizona and came back with the rest of our things and the '65 Valiant. We stole a water heater and a stove already plumbed for propane from a state-protected abandoned home. Now we were Mexican outlaws!

The owner of the property where we lived, Trinidad Fiembres, was also the owner of a chain of Mexican Supermercados, centered in Tijuana, where he lived with his lovely extended family. It was a while before we met him as he was on a cruise to New Zealand. But shortly after he returned, the entire family came down from Tijuana to the caravan where he had a substantial house with lots of bamboo privacy. The rest of the caravan had two other houses, ours and one for the caretakers. I don't remember if there were two or three married children of Don Trinidad, but they were our age with children also of compatible age with ours. All of them loved the blond hair of our contributions.

At some point during our first visit, Rachael did her nude dancing. Soon all the children, parents and grandparents at the compound were chuckling over the little naked blond fairy. Her nudity turned out to be another bonding factor with the Fiembres family, and while we never got to know them as well as we would have liked, they held us in high regard and for the remaining two years they let our rent fall with the peso which meant a drop in what we paid them from $120 to $30. That was a huge help for us. The Fiembres were patrician in their essence, but in the very best sense of that. It was memorable meeting them and their oceanside property was our home for about two and one-half years.

The rest of the caravan consisted of about a dozen getaway trailers, including no more than a couple of bigger ones. Most were built for four or fewer. Several had homey seaside decorations and or were sheltered with bamboo. The owners who came down from California, liked to party and soon became our friends. Mostly they came down for the weekends, and having worked till 1 p.m. Saturday with my week's pay in my pocket, I was in the mood to party. Not that our partying was that expensive, but often our friends understood that we had less than they and were quite generous.

We had campfires, drunken chats, occasional pot and taped music. It was a very social time for us, and my work sort of spread us into the Mexican community as well. Un Chavo, se llama Bonifacio, nos invito' a su casa. Boni was a kindly, but strange man, with a nice car he had brought back from California. Only he never drove it except on this Sunday, when we visited him. His $17-$23 per week pay didn't provide him with money for gasoline. I remember he lived alone in a very large house, typically Mexican, with lots of bare walls and concrete. I believe that he ended up with it after much of his family died or left.

We also spent an afternoon at a child's birthday party attended by parents I knew from work. Marsha told me that as soon as the men went outdoors, the women quickly pulled out cigarettes and passed them around in a kind of rebellion that Marsha thought they very much enjoyed. Cigarette smoking was an important social expression there; most people participated and there was always open-ended sharing.

Chuy was one of my best friends at work, a very smart, energized and intuitive man who everyone at the nursery called on when they needed help. He and his wife owned a washing machine and had electricity – Boni didn't – would we pay them to do our laundry? Yes, we would. And wasn't that the way it was in all colonial relationships? We were now upperclass Mexican outlaws. From work, just downstream from the Rancho, I could see women washing their clothes in the creek, then pounding and drying them on the rocks.

I remember an American woman we met who had a house a few miles from our caravan. She was talking about her maid who was paid $50 for two full days of work, but it took her one and one-half days riding on buses and a half mile walk to be there for the work. Our new acquaintance said she was amazed that anyone would go through all that for this job. I said to myself, un huh, she's getting $50 for three and one-half days and the workers at the nursery get $17-$23 for five and one-half days, dunh. Mexico doesn't really hide the expression of its social- and economic-class system. At $100 a week, we arrived on a certain plateau above the peons, but below our American with her maid, and far below the peak of patron inhabited by Jorge and Trinidad.

Once the house got the water heater for shower and dishes, we were very comfortable there. It was about 900 square feet with a big picture window facing the sea. Arena reminded me that I had to cut through the wall and frame in a door to the patio we built on the oceanside. We cooked, dined and did our reading together there. We bought brick for our patio and landscaped it also. We faced the Pacific and the setting sun. At spring tide, the ocean was only about 200 yards away, though we were on a bank about six feet above the pure sand beach. At low tide the surf was about 600 yards distant.

The ocean was cold there. I was the only one who swam in it, but never for long. Once a week we saw a cruise ship glide by on the way to Ensenada then back to LA. Plenty of debris washed up on the beach from their passing. We also just made out the puffing and chugging of gray whales when they migrated by. They left no debris. Once a dead sea elephant washed up on our shore. It must have weighed 1,500 pounds. It eventually just stank, rotted down into the sand. The tide didn't get it till the carcass was just a pile, two or three weeks later.

During the week while I worked, Marsha and the children would often hike to one rocky headland or the other. In those places were sea anemones, small octopus, and other inhabitants of the tide. During the week our entire three-mile beach was often empty of humans save for the caretakers and "the Muellers." On this beach occurred the famous Heath-Graham foot race. One fine, full-moon night, we gathered spawning grunions for the skillet. One fine sunny day we bought fresh-caught halibut from a Mexican fisherman who had trailered his large wooden dory onto our beach, then launched it in the surf. There was also fresh seafood to be found at the seafood market in Ensenada. Ensenada was about an hour south of km 71, but gasoline was cheap. We drove there about once every month or two. We went mostly for the fish market and shopping, but also for its natural beauty.

Driving south we passed several small canyons that as they opened and widened into the Pacific, were cloaked in strikingly large palm trees. Up canyon where the water was not salted by the ocean, there were little settlements whose size was determined by the amount of freshwater flowing from their canyon's small portion of the coastal range ten miles east and 3,000 feet higher. The chaparral between these canyons was clothed in xerophytes that lived on 7 inches annual rainfall.

Approaching Ensenada, the highway followed a ridge, hundreds of feet above the ocean. The descent revealed Ensenada's large circular harbor. I think it's the only real harbor the entire length of Baja. We always went on Sunday with the cash from my Saturday payday. Ensenada had about 120,000 citizens in 1985, and so there were a couple of good stores, both a Fiembres' Calimax grocery and a sort of early Walmart. I remember shopping there for Christmas presents, notably clothing and a large "transformer" for $20 and He-Man figures for $6.50. We always purchased as much fresh fish and shellfish as we would eat over the next week and carried them back in a cooler. That would be it for Ensenada, except that we must have walked around some.

I remember standing outside a popular American drinking hangout, Papa's, perhaps. It was a true hangout, people were loudly drunk and a few leaning over the second-floor railing, perhaps to escape the roar from inside. Marsha and I enjoyed drinking but only at km 71. One time we drove past the city and back up and around the bay to a place called La Bufadora. It was near the tip of the little archipelago and uplifted some 50 feet above the ocean tide. There was a kind of chute that allowed upward surging water to push a column of air through a kind a natural bassoon. Every few minutes La Bufadora uttered a kind of huge deep ocean sigh. Truly a sound to behold.

Most of our weekends we went north and did shopping in Tijuana. Tijuana, of course, in the states, has the reputation of being a crime-ridden dangerous city, but as "blond Mexicans" we had a very different opinion about it. It was the big city that had everything we needed at low prices. We bought tacos from street vendors and once in a while we went to a restaurant where we always very much enjoyed the food. Calimax was an excellent grocery with a delicatessen full of Mexican cheeses and meats. We never felt threatened there. We did not, however, participate in the town's culture, we never went to a movie there, attended no concerts or horse races, and did not picnic in any parks. I wish we had done those things. We did experience the Mexican culture simply by intermingling with the shopping public who found a way, every time, to feel the soft blond hair of the children.

In order to support the poorest in Mexico, the government subsidized two commodities, tortillas and dry beans. There was also a chain of government-supported stores with the name of Conasupo. We visited a couple of them down there; one was our closest food store in Ejido Primo Tapia. These small groceries lacked refrigeration and had the acrid smell of a third-world market. I am sure we only bought dry goods there. Of course, we were "middle class" and did not have to shop there. We did enjoy going to our tortilleria to purchase for about nine cents a pound, fresh, hot corn tortillas sold by the kilo. Most of the tortillerias I remember were in nondescript buildings about half the size of today's American dollar stores. There was a small customer area in front with a counter which blocked access to the interior. Three conveyor belts paralleled the three walls we faced. We saw an attendant load a large tub with about 50 pounds of moist masa. The tub had mechanical components that formed identical small balls of masa which passed under a roller on the first conveyor belt forming the uncooked tortilla. The advancing dough somehow turned the corner to the second belt which led through an electric oven the size of a small refrigerator lying on its side. From there the hot moist, perfectly cooked tortillas turned another corner and marched up to the counter where an assistant put down a square of waxed paper and then hand-mounted the hot tortillas until the scale said 1 kilogram. We paid the 19 cents, in pesos of course, and began eating the hot creations almost before getting back in the van. They were really good!

Entering the United States was a big part of our life in Mexico; we did that about once a month. Getting to Chula Vista could be an ordeal. I remember the trip as about 50 miles, but the first 40 miles up to Tijuana were on the old winding two-lane Federal Highway 1, also known as the free road. Then it was through busy Tijuana to the San Ysidro Crossing, the busiest on the entire border. It always took at least 20 minutes, but we sometimes waited an hour. We had one difficulty not shared by the other thousands who crossed each day. Our vehicles did not have current registration or insurance. Initially, the van had good plates, and later we were able to get 1986 plates on a trip to Chicago. The Valiant however, never had current registration. With a little blue and white paint, Arena advanced the year on the plate sticker. I'm not sure what we thought that would accomplish; you could tell it had been doctored. Once for some reason, we were taking the Valiant across instead of the van. That little car was 20 years old and reliable but smoked badly and visibly, exactly the kind of car Californians did not want to see. As we pulled in to the customs lane, we suddenly remembered Arena's license plate art. Sure enough, the customs agent walked around to the back of the car to check that plate. Sure enough she spent a little extra while in observation. She returned to her kiosk and without acknowledging anything, she did a bit of typing. She may have asked for my driver's license. But whatever, she waved us through. Registration crimes were a state offense and apparently the US government was not going to enforce state laws.

I cannot remember many groceries that we bought in Chula Vista except for the whole wheat flour or bread which we could not find in Mexico. Mostly, we got our mail and went to the nearby library, where with just the local PO box, we were literally allowed to check out 100 books at a time. Most of them were for the children, of course. We always checked out classic young adult books like Jim Kjelgaard's *Fire Hunter*, *Little Women*, or the *Little House on the Prairie* series.

Rachael and Heath, and certainly Graham, enjoyed listening to books on Arena's level. Arena attended the elementary school at La Mision for a month but that did not work out since they put this eleven-year-old into first grade. Who could blame them? Arena was tiny, about the same size as some of those first-graders. So for the nearly three years we were in Mexico, the books from Chula Vista, along with math and writing lessons from the parents, and the Pacific Ocean outside the door, turned out to be a pretty excellent early childhood education.

Two of the four Mueller students got PhD's, one got an MD, and the one who didn't finish college is now an executive with the Staples Corporation. They all later attended rural Oklahoma schools where academia was not held in high regard, but I digress from Mexico.

Although $100 a week stretched pretty far in Mexico, it disappeared fast in the USA. Fortunately, Marsha and I found that we could get $10 each for donating a pint of plasma. We did that several times in Chula Vista, and the $20 financed movies or even eating out. Once or twice we stayed overnight in the states, sleeping in the van. One of those times we pulled into a parking lot where a store had just been burglarized, the suspects driving off in a van. We got there and bedded down just before the police arrived. They gave us quite a bit of attention, but their flashlights soon determined that we were not the perpetrators. There was an awkward moment when they asked for insurance papers, but ultimately, they decided they did not want to impound our vehicle and strand the six of us. The luck of the innocent. A feeling began to arise in all of us. It was that we were relieved to return to our Mexican home. We had very much embraced our life there. It was home.

As I mentioned the work week was 48 hours. Cash was given out at noon on Saturdays when the week was completed. Work created a second life for me that was not really shareable with Marsha and the children. They experienced Mexico as a flow of life around them, but remained within an English circle. They were in exactly the same situation as many illegal immigrants around here, except that there were a lot more Mexicans who spoke English than there are Americans who speak Spanish.

We did have some social events with top to bottom of the local strata, not really that many. Sadly, I remain the only one in the family fluent in Spanish. No one at the Rancho spoke English but myself. Starting with my four years of high school Spanish from 20 years previous, I slowly at first, then more speedily began to learn the language. After a few months, I became fluent, though my vocabulary lagged behind since most conversation involved the nursery and its equipment. Anyone learning a new language and conversing with native speakers should realize that incorrect grammar and syntax, along with misunderstanding, can make one appear to be quite stupid in the new milieu.

On one of our breaks I picked up a stem of purslane, a plant which I had known for years and often consumed, and standing there munching the peppery succulent I stated, "Como me gusta comiendo verga larga." The Spanish for purslane is verdolaga which is similar in pronunciation but not the same. In English I had intended to say I enjoy eating purslane. What I actually said was I enjoy eating "big dick." It feels so good to chuckle along with your comrades when they are laughing hysterically. Doesn't it?

From the beginning, my job at work consisted of overseeing the watering and fertilization and to personally apply the soup of pesticides that carnations required in order to be sellable in the wholesale American market. Basically, I was spending about 20 hours a week fighting two pests, red spiders and thrips. My quick success in cleaning up the red spider impressed those in the know, including a man named Pedro who owned a neighboring ranch and occasionally visited. I ended up working for Pedro our last year in Mexico. I think there were about eight different flower growers just inland in valleys along our coast. Spider was a problem for all of them. I was the only one to recognize that carnations have very waxy stems and leaves so that anything sprayed on them immediately runs off.

For the spray to stick, a great deal of spreader-sticker (or detergent) was needed to get a film to form on the plant where the insects were. Once the soap was increased, I visualized that the very thin film would not hold much pesticide, so I also increased the strength of the pesticide solution. I engineered spray nozzles with which the operator could quickly walk a bed with up to four nozzles blasting the insecticide up from the bottom soaking everything. We needed far less water with my formulation and so we moved quickly, finished early, and killed more.

The only other serious pest of our carnations was the thrip. Well not really one thrip but the millions that survived perfectly well in the chaparral of the entire peninsula. I suppose their numbers must ebb and flow through some natural process. They were less numerous in winter, but we sprayed for them all year. Thrips fly in to the greenhouse attracted by the bright flowers. They only feed on the flowers, and in doing so they stain and wash out the colors, effectively ruining the flower for market. They also carry viruses and blights, a fact well known to US greenhouse tomato growers, a number of whom had to destroy their entire crop and add special netting to their houses.

The carnation flowers are perfect cups that stand vertically, so they are easy to spray and they hold the insecticide for a while as a miniature toxic lake. I also designed a wand for spraying the thrips, basically a simple boom the width of our beds. This did not lead to the hoped for success because poisons of that day were simply not very effective against thrips. I used four or five of the most toxic insecticides ever sprayed on the planet in hopes that a rotation would synergize their death. Not sure what did what and never was. I remember though, spraying methyl parathion and marking a spot I had sprayed more heavily. I went back the next day to look with a magnifier. Methyl parathion the pinnacle of the organo-phosphates was one of the two or three most toxic substances ever formulated. It had a bad reputation for killing California farmworkers. So when I followed up the next day, I found adult thrips still alive in the souped-up flower.

Using these pesticides, fighting for the perfection of florist carnations is most certainly involved with my bladder cancer which broke out 25 years later.

During my time at Alisitos I only saw Jorge two or three times. It happened that he was the PRI candidate for Diputado from Baja California Norte to the Mexican Congress in DF, the Distrito Federal, and that took all his time. In fact, the two young men we had met the day I was hired, were Guardaespaldas, bodyguards, required by Jorge's new status. I am not sure when I found that out. I had read that the PRI, the Institutional Revolutionary Party, always won, but Jorge was an avid campaigner as I found out when I drove up to his house early one morning to arrange for a fertilizer purchase from the states. I arrived shortly before he, the bodyguards, and his secretary boarded their van for a day on the campaign trail. I was invited to get in.

Thus began a remarkable day on the campaign trail in Tijuana. That day ranks high in my memory all these years later. I actually only remember three stops of the four of five that we made, and my weakness in understanding Spanish at that time further diluted the experience. Enough remains for my tale.

While none of the four I was with had any interest in talking to me, especially since what I did had nothing to do with their endeavors, I was treated respectfully by them and allowed to listen in and even participate in their chats as the day went on. One aspect of that is that the two bodyguards and I found the secretary to be very attractive. She was a Mexican redhead with freckles and had a positive kind of self-confident demeanor. Perhaps as a result, the van had a kind of pleasant banter to it throughout the morning. Our van was joined by an older cheaper model that was armed with a bullhorn loud speaker one never hears in the streets of America.

"Jorge Salceda Vargas para Diputado," it wailed over and over as we caravanned through the hilly streets of Tijuana.

Our first stop was at a barn/gymnasium somewhere on a plateau. The floor was raked dirt, but it did have basketball goals and soccer nets. There were no bleachers, but the attendees sat on long wooden benches or folding chairs. I remember most of them were older women. They arrived in the most formal of clothing and were a colorful crowd indeed. I stood in the back with Jorge's other employees and tried to understand his speech which was well delivered and well received. Soccer balls and volley balls were given away to whatever community organization was housed in the gym/community center. All this lasted quite a while, and afterwards Jorge sent away his staff so he could hobnob privately. This threw the bodyguards in a tither, but they consented that we needed to have lunch, so we drove a short distance to a café. I do clearly remember the conversation at lunch which revealed a kind of angst that Jorge was unhappy with them.

"He thinks we are made of iron like he is," they exclaimed. "We're only getting four hours of sleep a night and not one day off since the campaign started."

'Ni un Dia!" and the secretaria just nodded her pretty red head in agreement.

After a leisurely lunch, we returned and were able to retrieve Jorge without incident. No word from him about his lunch. At the next stop, Jorge pointedly tells me that I am to stay in the van until they return. I wondered if the association of my blond hair and beard might not be a campaign plus for him.

I was disappointed by my banishment and soon became bored enough to look around my surroundings. Outside nothing was going on and there were no compelling views, but on Jorge's seat was a very large zippered bank bag. I opened it. Inside was a three-inch thick wad of notes and a Colt .45-style pistol, big and heavy.

I hastily returned Jorge's things as they were before I touched them. The rest of the day was anticlimatic. Within about 10 minutes of our return to Jorge's house, he gave me the three hundred dollars needed to buy American fertilizer. He was quite glad that I could easily cross the border to pick this up, and the van was three-quarter ton so capable of carrying a lot. He threw in gas money and finally another $20, "Para la mordida" he told me.

By then I was familiar with the term from when we crossed in to Mexico with our load of belongings. We were held up by the Mexican border guards who were letting everyone else through. I think $30 is what we paid that time. La Mordida means "the bite!"

Days later when I drove up and bought the fertilizer, I got to see the incredible Monrovia Nursery and outdoor cut flower fields that made Alisitos' 25 acres seem tiny. With almost a ton of fertilizer inside, the van was riding a little low, so to speak. The guards were not going to miss that. They wanted $40, but by this time my Spanish was good enough to argue and invoke the name of the future diputado and they finally accepted the $20.

Nine years afterwards, we were in Oklahoma and I read of the assassination in Tijuana of Luis Donaldo Colosio. Colosio was a reformist politician, and the PRI nominee for president of Mexico. The analysis I read suggested that he was killed at close range by a bodyguard of a local politician. Holy Cow. I did not know Jorge that well, of course, but he was serious and businesslike with me while at the same time there was something he conveyed that made me feel comfortable in his presence.

It was easy to associate him with Pancho Villa and other legendary "traditional Mexican politicians." However, before feeding this image of my former employer, I note that Pancho Villa was a reformer and very popular with the people of northwest Mexico including Baja.

My favorite person in Mexico, Don Pedro, at whose nursery I worked at my last year there, was a very good friend of Jorge's and always spoke well of him.

Jorge's recent death brought forth correspondence of good things he had done for Tijuana dramatically enhanced by internet photos of his bier replete with foodstuffs, woven colors, glittery jewels, candles, mirrors, crystals, and little gleaming skeletons. It struck me as a kind of tear of culture.
Part of Jorge's legacy is the Fundacion Jorge Salceda Vargas, and they distribute food to the poor in Baja California Norte. How wonderful that Facebook has allowed me to correspond with Jorge's daughter Claudia, who is clearly a driving force like her father. From that group I learned that the greenhouses had burned down in a wildfire, probably for the best.

Claudia Salceda-Conijn wrote this about her father Jorge Salceda Vargas, my employer in Mexico:

    My father arrived in Rosarito with his family in 1957, when he was twelve years old. His parents, who brought them from Michoacan, believed there was opportunity on the frontier with California. Young Jorge worked three jobs as a boy forging his character through struggle and perseverance.

    Later he accepted important roles in the public. He was commissioner of his Ejido, Community delegate, agricultural leader in Baja California Norte, and Federal Legislator, Diputado, in Mexico City. His love for the land was a constant, and he initiated many rural projects, plantations for palms, ornamental trees, and for flower production at Ranch Alisitos, where he grew chrysanthemums, gladiolas, roses, and above all carnations.

    In the year 2000, he established a nursery which used an innovative recycled water system to grow palms. Before his too early death, he was spearheading Rancho Palmar Paradise, a center of nature and entertainment on the coast three miles south of Alisitos.

His death came suddenly in 2009, shocking his children and depriving us of his attentive guidance and wise advice given with much love and pride. My father completed in the highest degree his role as patriarch, good citizen and natural leader for many valuable projects. Rancho Palmar Paradise held his most tender affection, and he felt like it was our personal garden.

Not too long after my day with him, Jorge leased Alisitos to a management company. I did not care for the new manager, Carlos, but things stayed good except that my salary, now paid in pesos began falling with one of the many serious periods of inflation in Mexico. The peso went from 250 to the dollar to 3,000 to the dollar over a couple of years. Salary increases, of course lagged behind, and at one point my income dropped to only $67 a week. For Christmas, Carlos' firm gave all 75 or so employees a nice Mexican blanket; wish I had it now.

At the end of spring we were invited to visit Chicago, all expenses paid. The Heasleys of course wanted to see us. First, they had to send a couple of hundred dollars for us to buy new tires, one item that Mexico did not sell cheaply. We left in the summer and were gone two months. When we returned, Carlos was surprised; he thought we were gone for good. Shortly after that, I sought and was given permission to leave Alisitos to work for Rancho Las Flores, owned by the young and intelligent Pedro, who had always liked my work.

Pedro restored me to my $100 and paid me in dollars. I enjoyed working there more than Alisitos because of his daily presence, but by the fall of 1987, Marsha and I agreed that there was no real possibility that I would ever make much more money, nor was there any possible promotion for me with Pedro or anyone else. The children were still thriving, but Graham was now seven, and we had two children who would normally be in school.

Our key to a move back was Marsha's RN license. We would have returned to New Mexico, but unfortunately, in our favorite state, her license had gone inactive and to work she would have had to take refresher courses for eight weeks and pay for them as well. Ironically, she was quite employable in any other state. She was considering a position in godawful Needles, California, when one Saturday I came home and ran into a loquacious tourist named Carl Branscum.

Finding out we lived in Mexico, he inquired about whether he could not also live cheaply in Mexico. Carl was older, about sixty. He and his wife lived in Muskogee, Oklahoma. When I told him that we were leaving Mexico as soon as we could find a job for Marsha, he told us he owned an 80-acre farm near his Muskogee home and that he wanted us to live there. Marsha, he said, could get a job at the well-regarded Muskogee VA hospital. We knew a little of Oklahoma because we had driven through it often. This move to Carl Branscum's farm house, incredibly, is exactly what came to pass, and two months later we moved to a new home 2,200 miles east of our ocean.

This is a deciduous azalea, one of a number of cultivars called "Flame Azaleas" This plant flowered more spectacularly every year I was with it and remained attractive after flowering and through the nudity of winter.

This is mostly rye, from just a few years ago. I did my real learning about growing and harvesting small grains back in 1975 when I grew wheat, rye, and triticale in the manner described by F.H. King in Farmers of 40 Centuries. Grains are planted in clumps and harvested with a sickle. To thresh the grain, I made some numchuks from two short, hard branches looped together with leather twine. The numchuks seemed like a deadly weapon. I whacked the seed heads onto a sheet, and Marsha I then winnowed the grain by swinging the sheet on a windy day. You could see the chaff blowing away and the seeds were clean in no time.

We liked amaryllis and did well by them. The bulbs should never be tossed after blooming because they are easy to maintain all year and keep getting bigger and prettier.

22

Oklahoma 1987

The last couple of months before we left Mexico were bittersweet. Our friends from km 71 bought a cargo ship hawser we had rescued from the surf months earlier; the $200 a big help for our move. Then Fernando, my first boss, came to visit and stole the $200. We were so desperate that I got back in touch with this same thief and sold him the Valiant for $45. We actually made only one capital purchase while in Mexico. That was a dining table with six chairs. Its base was a gigantic burl of Mexican cypress with a pegleg to balance the massive round top. We spent $190 for it.

It was heavy, and along with books and what all else packed in the trailer put a quite a load on the van. Actually, I remember the camper and the van as just being full. I believe I put 50 or 60 pounds of air in the tires. In addition to the six of us, there were four cats that had become the beloved pet part of our lives. We went to some effort get the shots and vaccination records required for their migration to the United States. Then suddenly everything was packed and Mexico was ending.

The van slowly pulled out of our sandy drive lumbering like a walrus. I had decided that with the weight we should take the tollway all the way into Tijuana because our free road, Highway 1, had lots of traffic and uphill climbs and was a two-lane roadway. The toll road, 1D, I knew to be level along the north coast. I had only taken it once and that was from Tijuana south to km71. I had no idea of the peril awaiting us upon arrival in Tijuana. The north end of the tollway sits atop a hillside several hundreds of feet above the city below. As we rounded the crest, we had a great view of the sprawling city beneath, but our roadway was a very long and very steep drop into the city. I quickly discovered that though the van's brakes were functioning normally, on this incline I could not stop our momentum.

I remember 35 miles per hour at some point in the throttle down. I had dropped the transmission to second and was wondering about first as we approached a broad intersection at the bottom of the hill. I remember that there were three lanes coming down and we were crossing four lanes. There was a traffic light, red unfortunately, that we were running at 25-30 mph with minimal steering control. It was hang-on time. Of course, none of our children were belted. Our front seatbelts were laptop only, no shoulder harness, better to die than have that save your life. Our uncontrolled crossing was successful; perhaps other drivers saw or heard our frantic horn. We shared a huge sigh of relief, and on more level roadways, the brakes at least would stop our laden tandem.

Here is the cypress wood table today in Arena's dining room. We still have all six chairs as well.

Our hearts were still pounding because from the bottom and heart of Tijuana where we were, the San Ysidro crossing was very close. Within the blink of an eye we were slowing for the congestion at the border. We had crossed enough times so that we could correlate from our distance from la mera frontera where we would have to say something to a US border guard about our laden tandem. We did not want to be ordered to "secondary," which had happened to Marsha and I fourteen years earlier at the Ciudad Juarez crossing. That was bad; this with the children and the packing would be exhausting and harrowing.

I am now convinced that the four blond, healthy, tanned children were like a key to all doors. At the last minute before we pulled up to inspection, we discovered that while we had the permit for the cats, we had no idea where it was. Marsha, Arena, Graham could not find it. Maybe it was in a drawer in the jammed-up camper, who knew; stopping to search was a sure trip to hell.

And when we pulled in beside the kiosk and stopped, the kittens were in a cage directly behind my shoulders, visible I thought to the officer, who asked if we had anything to declare.

I responded, no, but that we had been living in Mexico for three years and were returning to the USA with all our possessions. He did a quick glance at my license and into the van and then asked if we had any pets. I took a deep sucking breath and said, no. He waved us on through, and from there we drove to one of our California friends who had a home near Interstate 8, our route to Phoenix and points east.

After a poignant evening with our friends at their home in the foothills east of San Diego, we rested a bit then headed further east into the night. From the summit of the Coast Range on Interstate 8, we descended to the bottom of Laguna Salada. The highway here has steep parts, some even with emergency cut-offs, but nothing as bad as the Tijuana hill. We held our speed to 35 mph and made the bottom safely.

Of somewhat more concern was the fact that once we were settled and driving on flat ground, our fully functioning temperature gauge was pegged at or just below the line for full red hot. It stayed this way for the entire trip, even though we did not drive in the heat of the day. The third night around 11 p.m., we crested a ridge and saw the lights of Las Cruces, New Mexico, below. That prompted a discussion about which of three possible routes we could take from there. I somehow chose clearly the worst of the three.

The van when it was new had driven several times over the 5,800-foot pass at Cloudcroft, but it sure wasn't loaded like now. In the still moonlit night, we could see the long uphill which led to the real steep part of the highway. I built up what speed I could, and it wasn't long before I had the van floored. The temperature was at full hot and yet our speed slowly slipped away. Maybe we peaked at sixty-five, then fifty, forty-five. I never lifted my foot from the floored accelerator; at 40 mph, I pulled the transmission into second. The 302-engine kept pulling, but our speed continued to bleed off. At 25 mph, I dropped the transmission into first gear, and we crawled up the last quarter mile to the summit. The motor was roaring but in first gear we were moving slowly at the end. Then the right rear tire of the camper exploded, and we pulled off at a safe spot on the very crest of the pass.

We slept where we were as best we could. There was little traffic as we were now well into the morning hours. After we stopped the motor, its console kept pinging with heat, for long slow minutes, still pinging, ping, ping, ping. In the morning I got out and changed the camper tire. The van didn't start, but with some pounding on the solenoid and a bit of a scare, it kicked right over, and we were once again on the road.

Turning off I-40 the next day, we took US 60 up to Canadian, my mother's childhood home, where my cousin Tommy and my grandmother resided. Tommy ran the family lumberyard and was very well off. He owned a 75-acre mansion on the Canadian River and a house in town, which he put at our disposal. It was a house his 21-year-old son had shot himself in a few months before. We ended up staying there just one night as I recall. Tommy was unavailable, which was too bad because I have now not seen Tommy since 1973. We did spend lots of time with Gran, who at 93, was still living in a little house in town.

She fixed us cornbread fried in honey, and my young children at least got to meet a great-grandparent. I met two of mine when they were in their early nineties. I remember being startled how a person could look that old. Perhaps our children carry the same kind of memory. We also got to visit with my younger cousin Jane and her son Seth.

Sand's grandmother

Then on to Oklahoma. By phone or letter, or I don't know how, we had agreed to rent the farmhouse for $250 and talk about purchase later. Our first arrival was therefore at the Branscum house in Muskogee, where we paid the rent, picked up the keys and directions to our new home, which turned out to be 23 miles, about five of those on dirt roads, south and west of Muskogee, where Marsha would be employed. The house sat on the high point of a moderately sloping 82 acres. Fifty of those acres were leased out and cultivated. This was unusual; there weren't many farmers in this part of Oklahoma who were still growing crops on upland soil. Land like this had been cultivated to grow cotton in the early years of the century, but the soil had worn out quickly and was now almost all maintained as pasture.

Branscum's leased acres had harvests of wheat, soybeans, and milo while we were there. The other 30 acres included a couple around the house and barn, the 900-foot drive, a three-acre pond and twenty-seven grassy acres which was cut and baled by a local farmer once or twice a year. He probably sold it to the dairyman three miles distant. Council Hill was definitely Oklahoma prairie. It had nice looking pastures, plenty of hay, a nearby dairy, plagues of junipers and broom sedge, deep bar ditches lined with pink or yellow heritage roses and/or elderberry. Our farm had a long unobstructed view north, all downhill and about a mile to a set of railroad tracks.

The first few months we were there a train still passed on those tracks, but after that, nothing. The curious thing is that even though the train was a mile away, it made our house vibrate, unmistakably vibrate. What was the soil structure up to?

Town was two miles west on the limestone county line road. There was a decent-looking school with grades 7-12, several crummy or abandoned type buildings, a cluster or two of trailer homes, a couple of street signs, a welding shop and equipment yard, and 123 residents. Council Hill was named after some powwow with the Indians, which none of the Anglo residents I met knew much about. We had no Indian friends that I knew of when we lived there, but I am sure Creek children went to school with mine and were friends of theirs. Black people lived seven miles north in Boynton, one of the original black settlements following the land rushes. Council Hill was in the heart of the Creek Nation and I know they maintained and still maintain their culture all around that country. I did not get a Creek Indian friend until much later.

The farmhouse was wood-framed with a cistern at one end and crawl space at the other. It had natural gas service and adequate heating and air conditioning and seemed big to us after Mexico. After just a few days of unpacking and organizing, we left the trailer and took the van to Chicago. Marsha's parents had bankrolled a big portion of our return, and now were gifting us more by having us drive up for a few weeks before Marsha's job started.

After our visit, they sent us down with Marsha's sister's car a worn-out 70's small Ford sedan. It lasted about three months, but by then, we were able to pick up something a little better. This left me to stay at home on the farm with Heath and Rachael, who were to begin together in kindergarten and first after two years home alone with their father. It was a half mile from our house down to County Line Road, where Graham and Arena caught the bus. From the nearby Midway High School, where Arena attended, first-grader Graham took a second bus to his elementary school ten miles down the road in Hitchita, Oklahoma, population 266. It all seemed to work and I had the freedom to build an attached greenhouse and till a substantial garden.

For the next two years I was "house husband." I found that to be a hard job, and putting out nice meals was not my strong suit. Marsha from the start had difficult hours. I think she was on rotating shift at first, but eventually went to straight nights. In early 1990, we made an offer to buy Carl's farm, offering him the value an appraiser we hired had put on it: eighty-two acres, city water, gas, three-bedroom house, 30-by-40-foot barn, 900-foot gravel driveway, $49,000.

Carl wanted sixty thousand and I don't blame him. After that we moved to a house in nearby Wainwright, and out of the blue, I got a job 64 miles away in Tulsa.

My employer, A New Leaf, was a greenhouse and sheltered workshop. They had had the chutzpa to apply for a landscape maintenance contract at Tulsa's brand new state college campus, the University Center at Tulsa, abbreviated UCT and pronounced "ucat." I was hired for the tiny sum of $6.50 an hour to take a crew of developmentally disabled adults to this "jewel" of a campus and keep it mown, weed-eaten, and irrigated. I hastily learned how to do that. We were also per the contract, to apply pesticides and fertilizers as needed.

Actually, A New Leaf got the contract because the state had a clause supporting 501-3C organizations with state bids. In fact, the only thing my crew members could do was hand weed and pick up trash. The facilities manager, Mr. Lew Donnell, was not pleased, and it was perfectly clear that the contract would not be renewed; yet somehow he liked me, and after the end of the contract in July, he hired me as grounds supervisor. My pay doubled to $26,000 and I had state benefits.

It was a stunning validation of something. I never should have been hired for that job, which a lot of Oklahoma State Aggies would love to have gotten. Be that as it may, UCT was the site of my greatest public success, which turned out to be more than I could have imagined.

Never saw a trunk this yellow. It's on a maple.

My gardening friend Robert gave me a couple of Japanese maple trees he had dug up as weeds. Of course, they were no special hybrid but both were shapely trees with pretty leaves

The happiest of Heucheras. These were so pretty in my shady bed that I just kept wanting them to grow faster.

This photo isn't doing justice to the 12' tall, 20' across saucer magnolia.

23

My Six Years at UCT

UCT stood for University Center at Tulsa. It was a commuter college with large parking lots and quite a few too small parking lot islands. These had sickly ash trees in them, sickly because the campus soil was damaged. There were 20 acres of modern irrigation which watered a number of flower beds, some of which had the "founder oaks" in them. Most of the irrigation was for Bermuda grass turf. I was hardly aware of it when I began, but the total campus land was 200 acres, a number of which became very interesting later on, but in August 1990, when I began, the challenge was to keep the installation they had completed a year and a half previously in good condition.

The site work and plantings had all been done before I arrived, but it was obvious that a lot of clay subsoil seemed to have passed for topsoil. Everything was tight, depleted, and poorly drained. All the soil on this landscaped part of the campus had been pushed with a bulldozer. The landscape architect responsible, along with the contractor, for the success of the project was getting pretty worried. The founder oaks were impressive 25-foot tall specimens and lined three main entries to the first large classroom and administration building. These trees were grown in the nursery ground to that height, then planted at UCT along with everything else in a hurry in 1988. They were pin oaks, which are marginal in Tulsa and require an acidic well-drained woodland soil. These were set in special beds lining the entries, and the beds were as poorly drained, tight, and alkaline as the lawns were; damaged clay covered with cheap mulch.

All the founder oaks foundered. They were going to force the poor contractor to buy 25 more of these monsters, but after they brought one tree, I told them to stop, that the beds had to be rebuilt before they could be replanted. Months later, we put in redbuds and white buds which were much prettier, and we did this easily in house once then-Facilities Manager Lew Donnell and I agreed on what we wanted. The plantings came to about one-sixth the cost of the original failure. I did manage to keep the parking lot ashes and a few other oaks and pines alive, so only the twenty-five 25-foot founders had to be pulled out and hauled away.

My top priority at the beginning was to keep the campus lawns looking good. I don't think they would have hired me if they knew I hadn't mowed lawns since I was sixteen and had no turf or arboriculture experience either. Fortunately, another supervisor had bought a nice 52-inch, walk-behind, zero-turn mower that was able to cut all that we were mowing when I took over. I had one full-time helper from the beginning.

Turf, like all aspects of horticulture, is a rich subject area which combines science, continual observation of soil, weather, timely and adequate responses. When that knowledge is used and the physics accomplished successfully, there will be healthy vigorous plants. After that, in strategic places the turf is removed to allow for trees, shrubs, and flowers to create beautiful landscapes. It was fortunate for me that frequent high cutting, regular watering, and fertilization can usually be counted on to grow good Bermuda grass turf. I had the wonderful opportunity to attend an irrigation workshop given by the Toro Corporation at the Shangri La Resort. It was phenomenal, and at the end of a weekend with food and a hotel room provided, I left with complete and expert knowledge of all aspects of this important part of professional horticulture.

The following summer I made, in house, my one and only herbicide application, apart from Roundup spot spraying. It burned a little, not too bad; I used some kind of three-in-one expensive chemical, and by the fall I was thought to be competent by all concerned. Soon after that I began buying organic fertilizers, and no synthetic products were used on the turf until I was dismissed.

After the mowing needs slowed down, I began hauling manure from the Tulsa Zoo. We started with just the state pickup and then added a trailer. At the same time, money and equipment, and new areas of care, kept flowing my way. I soon had a tractor with a loader; a 21-horsepower, 72-inch Grasshopper mower; a four-wheel steering, 48-inch Kubota riding mower; a decent roto–tiller; and professional grade Weed Eaters and chainsaw. I was allowed to hire a third part-time employee, who after a while became full-time. With this mechanical support, I extended our work into the edges of the campus.

Before I arrived, one of UCT's vice presidents, Dr. Charles Evans, had followed up on an interest in historic trees and arranged for the Tulsa Park Department to grow some of them for us. I think most were seed grown, and when I got them in 1991, they were of decent size and quality. Planting them on campus opened a doorway for me to step up to a level of importance rare for a horticulturist and even rarer for a horticulturist who had never taken a single class in horticulture.

Charles and I decided we should turn the historic tree planting into a public event for the university. At this same time, it was said that the mayor of Tulsa, Mr. Rodger Randle, was thinking that he would rather be president of our new burgeoning campus than mayor of Tulsa. So he came to our ceremony and made a few remarks. I got the opportunity to deliver to the crowd of about fifty, an eleven-minute speech. I was advised that five minutes was kind of long. Mayor Randle was impressed. The recording is lost but the written version follows. The speech is framed around eight historic trees we had just planted. As I referred to each tree in turn an A New Leaf client, a former contractor to the state, raced out to that tree placing its attractive historical signage. Our site was large with a central pond so this visual setting added much to the speeches and ceremony. Here's the speech, given to the mayor and about fifty attendees including press.

Historic Tree Dedication Speech
March 12, 1992

UCT Grounds Supervisor Sandy Mueller

First of all, I would like to acknowledge Mr. Pat Standingbear. Pat grew these trees on for the past two years at the Tulsa County Parks Department. Thank you very much, Pat. (Applause). I'd like to briefly mention that the University Center is sponsoring a pair of community gardens this spring. One is for the residents of the Sunset Plaza Apartments, and the other is for the Tulsa Community Correctional Center.

Last month, I had the pleasure of attending Mayor Randle's Urban Forestry Luncheon, and it was very exhilarating for me to be among so many people who were planting trees and involved in the environment. So perhaps, I should feel the same today, except that these are

"Historic Trees" we have planted. Well, not really historic in themselves, of course, but descendants of historic trees. As such they are part of a greater history, that is, the history of the forest from which they came.

And over 400 years ago when the parent of the little water oak we planted across the street sprouted, it was the newest member of the greatest forest ever to exist in the temperate zone since before the dawn of men. That forest extended for over a million square miles. Marching north from where our little seedling's parent was growing near Chesapeake Bay was an unbroken stand of the magnificent eastern white pine. When John Cabot became the first European captain to sail this coast, he claimed it all for England, north all the way till it butted up with the French claim for Canada. Seeing these trees, he was surely amazed, because in all of Europe there were no mast trees, straight-trunked evergreens taller than 100 feet. These American pine were everywhere, 200 feet up to 265 feet tall, and in a couple of generations, ships went from the size of his ship or the Santa Maria, which could have floated here on our pond, to vessels which appear gigantic to us even today. Not only were they tall, but white pine masts were lightweight and limber; of all the species ever used for masts none could compare with white pine. Cabot's claim, and the trees on it, would guarantee English naval superiority for the next 200 years.

However, the English soon became alarmed at the rate at which the colonial settlers were girdling and burning these trees for their farms. Therefore, the King sent his agents through the forest marking the finest of the pine with an "M" to reserve them for the masts of the Royal Navy. This aggravated the colonists, who often cut down these trees first of all. This aggravation and the great many others, eventually led to war. And in that war for the first time in over 100 years, the English suffered defeats in ship-to-ship duels.

John Paul Jones wrote before he left port in his first command, the Ranger, "I am sailing with three of the tallest masts ever put on a sailing vessel."

And thirty-five years later when a second war was fought, the USS Constitution with white pine masts and a hull made from the heartwood of southern live oaks was so incredibly powerful that English cannonballs could not pierce her, hence her famous nickname, "Old Ironsides."

During that war, the Revolutionary War, a small sycamore overlooked Washington's headquarters at the 1778 battle of White Plains. And about the time that sycamore was first growing, Daniel Boone was crossing the Appalachian Mountains into Kentucky. What he saw there was astounding to him: mighty beech, beautiful beech, walnut, oak, elm, chestnuts forming a 200-foot canopy of deciduous hardwoods. But unlike the eastern forest which was dense, thick, and dark dominated by pine, this forest was open. Boone described it as a riding forest where a man could travel, even hunt, on horseback. Imagine the beauty of such a place. These trees were twice as tall and four times as broad as the trees we see today. Consider the life supported by this forest. Remember the legends of the sky-darkening passenger pigeons. Consider the deer, the forest buffalo, the many other songbirds and eagles. Consider the soil microorganisms and mellowing effect on the climate, how the hardest driving rain would fall gently to the littered forest floor and be released in clean, clear streams and springs.

In 1830, Andrew Jackson lovingly planted the parent of our little magnolia on the White House lawn in memory of his wife Rachel. This now stately tree is pictured on our $20 bill. But by 1830, the great forest was already largely destroyed, mostly girdled and burned, as I mentioned, to make clearings for farms that all too often quickly eroded away. An astonishing number of these trees were used to pave the National Road where it wound through the swampland of Ohio and Indiana into the fertile north of Illinois.

In 1832, Washington Irving traveled through Oklahoma, very near our campus. He wrote, "We were overshadowed by lofty trees with straight smooth trunks like stately columns and as the glancing rays of the sun shone through the transparent leaves, tinted with the many hues of autumn, I was reminded of the effect of sunshine among the stained glass windows and clustering columns of a gothic cathedral."

Cathedral was a word many of the early naturalists use to describe our original forest. What a pity that we today cannot enter this cathedral because there no longer exists on our planet a forest the equal of this one.

In 1881 as part of the centennial celebration of the victory at Yorktown, a locust was planted at Carter's Grove in Virginia. In that same year, Charles Darwin published his second most famous scientific work, a treatise of the activities and benefits of earthworms. And just a few short years after that, Oklahoma was opened up to non-Indian settlers, and soon after logging of the western Ozark forest began, dooming the last stand of the original eastern forest.

In 1912, when the city of Tokyo, Japan, sent the first of thousands of beautiful blooming cherry trees for planting at the Tidal Basin and Jefferson Memorial, logging began in the Rocky Mountains. In the western shortgrass prairie, the land was plowed to plant grain in order to make up for failures in the crops further east, and the boll weevil was enjoying its third summer in the southern cotton fields, where at that time fully 15 percent of the nation's hard currency was earned.

In 1930 when Thomas Jefferson's design of 150 years earlier was finally realized with the planting in stately columns of the beautiful catalpa tree at Williamsburg, Virginia, the cotton crop failed completely and in my home state of New Mexico the 200-foot virgin ponderosa pine were being rapidly logged out. The early grain yields of the Western great plains were plummeting from drought and the Dust Bowl was building. The nation entered a great depression caused less by the vagaries of Wall Street than the exhaustion of our soil and forest resources, the true foundation of our civilization. In that same year, mycorrhizal association was discovered and described. It is a symbiotic relationship between the feeder roots of plants, especially trees and fungi, which fuse together forming a new kind of tissue which extends directly into living soil benefitting plant and fungus. Where mycorrhizal association is strong, plants show a remarkable resistance to pests and diseases. During that same period of time, Sir Albert Howard was developing the Indore method of composting manures, which is the method we are using on the campus today.

In 1937, the US Department of Agriculture published its most famous yearbook, *Soils and Men*. That yearbook catalogued the devastation of our soil and forest resources and provided a blue print for a great conservation movement which began at that time. This movement was characterized when the Works Progress Administration planted sycamore trees on Ellis Island and around the Statue of Liberty.

Unfortunately, this conservation movement came to an abrupt end with the advent of World War II, and after the war was over, the effort was not continued, but rather fossil fuels continued to be burned in the former ammunition plants to make chemical fertilizers. These fertilizers and their sister products, the herbicides, fungicides, and pesticides, make up the most significant aspects of the Green Revolution, which has largely made up the agricultural deficits of the Thirties. However, these products damage the soil wherever they are used; fungicides put a quick end to mycorrhizal association; pesticides kill birds, fish, beneficial insects, even people as well as the target pest. NPK burns out organic matter in the soil, and herbicides destroy useful variety to promote monoculture.

Therefore, it is the intention of the University Center to no longer use any of these products on our grounds, but rather to work toward the improvement of the fertility of our soil with the understanding that a plant grown in healthy soil will take care of itself.

In conclusion, a great many of us feel the need for conservation in our hearts; a great many of us practice conservation in our homes and gardens. We realize the terrible condition our planet is in today. Even as I speak, however, clear-cutting is still going on. The virgin redwoods and the last two percent of our virgin forests in Oregon and Washington are threatened, and loggers and logging companies are demanding the right to clear-cut these trees and send the raw logs to Japan. Surely, we must see that we have made great fortunes off these natural resources. Can now we not begin to replant and conserve so that one day our descendants may reenter the forest cathedral that we no longer have?

UCT was embarking on a second phase of construction larger than the first and to oversee the project had hired a top-notch facilities manager, my new boss, Jon Mercer. Jon and I had a slightly uneasy relationship, but he was basically a good man and a skilled engineer, also knowledgeable about grounds. He must have spilled his coffee upon hearing my pledge to use no chemicals on the grounds. On the other hand, all that was behind it was my own pledge, oh, and one little chemical I allowed, one advertised as and considered to be almost as safe as water, was a hormone disrupter, not a poison; its name was Roundup.

We never sprayed large areas with it, so our usage was not that high, but every sidewalk crack eventually became home for weeds, and some flowerbeds, particularly the ones I had not yet amended with our compost, were weedy enough to beg for a quick and dirty spraying rather than the hours of hard work required to do it the right way. Merited or not I claimed an organic campus and got positive attention from the media and my peers.

Within a month or two, Mayor Randle resigned from the city and became the new president of UCT. We reacquainted soon after he arrived and became friends. He became my number one cheerleader and introduced me to affluent movers and shakers, like Nancy Feldman, who was Chairman of the UCT board and a member of the Oklahoma

State Board of Regents. Both she and Rodger loved trees and the environment. Rodger invited me to his house to have lunch one day, and we talked about his landscaping. Later it was let out that I had permission to walk up to his office to see him whenever I wanted, something I did not realize at the time that unprecedented access to power by an underling inevitably leads to jealous reaction against him or her by the important subordinates.

The vice presidents had no intention of treating me better than the sweaty groundskeeper I was. Nor did my pay correspond to my high-level connections. I was really happy when I started at $26,000, and after five years merited $29,500. Looked at realistically it was a good, but modest state job, with a long commute. But in my ego world, it was like getting everything I wanted: wonderful comrades, plenty of help, lots of admiration, and connections into the heart of Tulsa, the nurseries, the parks and urban forester, the Oklahoma Tree Bank Foundation, "Up with Trees," the Urban and Community Forestry Council, Keep Oklahoma Beautiful, and beyond them, many wealthy supporters of the environment. Tulsans remember Sid Patterson, Tom Freeman, Joe Schulte, and Jon Kahre from Up with Trees, and from the city Joe Roberts and Mary Ann Summerfield. I became good friends with Mary Caffrey from Oklahoma City, who established and managed her inspired Tree Bank Foundation, and George Perkins, the University of Tulsa groundskeeper. I knew and admired them all.

We connected with a pre-release corrections center in nearby downtown and from them got a work crew of from three to eight inmates who we picked up in an old van my boss got us for this special task. For their full-time work, we paid like $13 a month per inmate. My staff was eventually expanded to three full-time state positions, plus myself. UCT was in the heart of a historic black neighborhood, and the closest residents to the campus were mostly black. Good then that my first two hires were strong and smart black men.

Gerry Wayne Ellis had recently gotten out of prison and we chatted on the sidewalk. Did I have an opening? No, but then soon after I did get a new hire, state job.

I saw Gerry again and said, "Let's fill out the paperwork."

Can you imagine being able to hire a state employee now without 30 days of ads, required interviews, background checks? Gerry was an awesome and charismatic man.

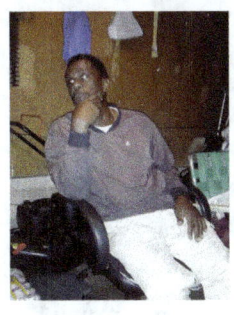

I then hired Jerry L. Jackson, who was quiet and cheerful, a perfect foil for Gerry and I. Later for a few months, I was able to hire Kent Smith, who looked like a choirboy with a touch of chewing tobacco. Kent brought a lot of previous landscape knowledge to our crew. I was the object of many jokes and even hysteria on the part of Gerry and Jerry, but my serious love of our landscape was taken on by these men who in turn loved the variety of our work, tractor, mowers, Weed Eaters, watering, growing, planting, living with the campus and seeing it evolve and grow strong.

We did a fabulous job. Our prison workers allowed us to set in many flowers, many potted plants, inside and out, sweep, the sidewalks, and get trash out of the parking lots. UCT started a student survey, and for the two years, I saw it, "buildings and grounds" pegged at the top of the survey, a full grade higher than any other department and high for grounds among all universities.

By this time the campus had built me a 40-by-60-foot steel building, a good one from the Mueller Company. I made sure they put the company logo with the Mueller name on it on the entry bay and on the door to my office. This wonderful shop, apparently named after me, was designed to allow me to add on an 18-by-40-foot greenhouse which cost the state only $1,000 and became a valuable part of my useful greenhouse knowledge and understanding.

It was first put to use growing the flowers I was most familiar with. We planted a lot of them. Our inmates made this extra beauty possible and freed the three of us to do other things all over the 200 acres.

We all ate lunch together in our building where we had a fridge, bathroom, everything we needed, plus a basketball court we built from spare state materials. Our inmates loved working at UCT, and everyone at their facility wanted to be assigned to UCT. I used to let spouses or family visit us for lunch. They were so appreciative because they missed their loved ones. I am sure it was against the rules, but we were able to do it anyway. We all knew and were friendly with the UCT police and their chief, Arlo Rudd. I don't think much escaped Arlo's gaze. He and I were about as different as two men could be, but he was scrupulously professional and we got along fine. He and

Gerry got along best, Indian lawman, Black con. Perhaps we had a reputation of being renegades, but we never had a scandal and the work we did was beyond reproach. It was by far the most joyful and rewarding job I ever held, and I am sure my mates felt the same way.

Simultaneous to all this was our massive Stage Two construction. All the remaining land from the bottom of the hillside to Greenwood Avenue was for an elegantly designed classroom, administration, and theater complex connected with a landscaped patio. Most of the traffic into the complex was designed to pass through the central patio, whose beds had to look good. So it ended up that I saved the day for those planting beds which had the potential to be much worse than the original beds put in for phase one.

I entered into the heart of a huge construction project. It was fascinating. My boss, Jon, was an engineer who had managed other large constructions for owners. He liked and respected me, but must have been as surprised as everyone at what Rodger said in the big pre-construction meeting for the final go ahead.

In front of Jon, the architect, the top foreman for the Manhattan Construction company, and quite a few other involved professionals, he said, "If Sandy Mueller has a disagreement or suggestion regarding the dirt work or landscaping of the project, he has the authority to stop work until the problem is resolved to his satisfaction."

Wow, I know they never heard anything like that before or since. I was the sultan of soil.

The design was for five large beds to be set into gorgeous pavers which extended some 300 feet uphill from the street to the theater. Several feet of limestone fill had been packed over a very poorly drained field, where once there had been a lot of ashes, to establish the grade for buildings and walkway. The architect had planned on twelve inadequate inches of topsoil for the beds, but a drawing error reduced that to an actual requirement of only five inches which was all that Manhattan intended to supply. So, I used my authority and Manhattan changed its schedule so that I could have two weeks with those beds.

My brothers and I did all the work in house. We trenched a drain which curved down from the highest bed through each lower and down to the street sewer. We did spec work with perforated pipe and gravel, and the 500 feet of line drained perfectly. We hand dug as much crap as we could out of each bed and hauled it off in our workhorse Cushman grounds vehicle. We replaced it all with many yards of the composted zoo manure we had made on campus. We had plenty because by this time the zoo was bringing the manure to us in their dumptrucks. The EPA didn't like it in Mohawk Park.

Unlike the beds for the first building that I had inherited, which are probably still bad to this day thirty years later, the new beds were excellent, and the Crimson King Norway maples planted in them grew like crazy and soon offered the perfect canopy while our department's flowers and big pots began to augment the design.

As the construction came to a close, followed by the grand opening, I continued work in the park and planted there and other places on campus some $23,000 worth of trees from an Small Business Administration grant program supporting nurserymen and landscape contractors. This grant program had run a number of years, and for Oklahoma, $250,000 was allocated each year and it stayed at that level for several years until the SBA ended the funding. The state was required to administer the grants and they created a council to read and judge the grant proposals and to be involved in the annual conference with hall, lunch, speaker, and elections.

Since I had gotten three different large grants, I became involved with the council and met leaders in non-profits, city employees, other government people, educators, and peers. So, for two years running 1995-1996 and 1996-1997, I was chairman of the Oklahoma Urban and Community Forestry Council. I guess that was the top rung of my climb on the horticultural status ladder. And it wasn't long until my connections and my "important job" opened up speaking engagements for me. I enjoyed and still do enjoy giving speeches. I spoke at several conferences, particularly the Horticulture Industries Show and the big Keep Oklahoma Beautiful annual conference. Most of my speeches were good ones. I gave several on the history of agriculture as described best by a book from 1954 by an Oklahoma county agent and a newspaper man entitled *Topsoil and Civilization*.

The end of my beloved state job came suddenly and wouldn't you just know it, at the height of my successes, which by their nature must have offended someone. Shortly after Christmas as 1996 began, my talented and fair boss, Jon Mercer was suddenly relieved of his job as facilities manager. Curiously, he was kept on in a small office far from the halls of power for one additional year at his full $54,000 salary. Even more curious was the promotion of his assistant, Jeff, who shockingly told me that Jon had been dismissed because he had made inappropriate sexual advances to Jeff's wife over Christmas. I am not sure if I believe that and wondered if Jon had been set up. I do not think he would have been kept on if he had done that.

Jeff was no intellect; his degree was in Sports Management. A year or so after I began, he was hired on as a janitor, then to chief janitor, then made assistant to Jon for the whole department. He was still in his early thirties, and in no way was he the equal of either of his predecessors. The three of us played golf together, but Jeff was never my type. Sometime that spring, I got a note from Dr. Vivian Clark, UCT director of Assessment and Research. She wanted to meet me and show me the results of her extensive student satisfaction survey. She wrote, "I wanted to meet the man whose area received the highest rating two years in a row. Not only did the area of buildings and grounds surpass every area at UCT, in terms of ratings, but was also significantly above the national average."

Meanwhile, Jeff did me no favors from the beginning, but began dismantling my authority and the Grounds Department as well. The first shock came when he terminated our worker release agreement with the state. The three, four, five, six, seven, or eight additional helpers for my department were gone, just like that.

Shortly afterward, he demanded that the three of us work separately, each responsible for our own third of the overall work. He said he couldn't manage the Grounds Department unless he knew who was responsible for what. I made a formal protest through Human Resources over that demand. I also found out about this time that the Grounds Department was to be eliminated beginning July 1, 1996, and the grounds work would be contracted. I no longer had a job to defend. There was a hearing with a new vice president I did not even know, the upshot of which was that either I did what Jeff told me or that I would be fired at his discretion.

A few weeks later that is what happened. Nancy Feldman, on the Board of Regents and a great fan of mine, tried to save me a position. Another meeting with three vice presidents, two of whom did not like me, and the third, a former backer, told me that little could be gained by trying to maintain any position with the university.

I had some advanced warning though. In late fall of 1995, Rodger's up-and-coming young assistant announced with great fanfare that the Helmerich Foundation had donated $25,000 for much needed landscaping on a new acre of lawn east of the new classroom. I had had no opportunity to improve this ground which was nasty, salty clay graded up to the base of the new building and compacted for the building's stability. It was the same shit dirt usually left for groundskeepers after the general contractors say goodbye. So I rented equipment and trenched in drainage, covered pipe and all. We also top dressed with our compost. The work came to $4,000. That got me into the paneled offices next to my friend Rodger. That money was not in my budget; no sir, it was not. I assumed that the grant money was for our department to implement ourselves or at least manage if it went to a contractor. It was a reprimand left hanging in the air.

No! That money was donated so that we could pay the Helmerich Foundation's choice for the contract. That contractor, Kingdom Landscape, was supposed to get the entire amount; they were not happy with my $4,000 bite. Since Kingdom Landscape was a company trading on its association with the Christian faith, I queried and found that their church was the same one long favored by the Walter Helmerich family. And then I saw, incredibly, in the Tulsa World, a newspaper article that this same church was putting on a dramatic play and the lead actors were Jomo Helmerich and my own new boss, Jeff, also a member of this same church.

Jomo was Patriarch Walter's son and heir apparent. I guess they were all praying to get rid of me, especially since I kept finding flaws, shortcomings, and mistakes on the part of the "Kingdom." They even had workers toiling without wearing a shirt. It was unprofessional. Didn't matter; they did their lousy little job and got paid for it. A silver birch indeed, in Tulsa, and a giant rock plopped on one of my sprinkler heads. Shortly before I was let go, I met the old man at an event at the Greenwood Cultural Center. At a dinner there, I had just been commended for all the landscaping I had done for them. Walter Helmerich gave me a weak paw and a sour dismissive expression. He probably knew about the $4,000. He also knew I was gone, and I didn't. I wonder if the contractor who replaced me attended that same Helmerich church. I believe that was Jeff who made that decision. Now 25 years later, the grounds of this campus look like they came out of a can. There are no more flowers.

The Golden barberry provides a luminous center for young azaleas. I never got tired of this plant or its purple brother in the back yard.

And in the back, in the dry fencerow neighborhood, the purple barberry thrived along with sparkleberry and a trio of hollies.

What a big fig tree, no figs. Behind is the grape arbor, 20 lbs. per year. I am a great proponent of growing grapes on large attractive arbors. I planted two one-gallon Fredonia grapes, and they grew to cover an area 30' by 40' within a few short years.

24

Historic Campus

During the Fall 1992-93 semester, administration somehow connected me with a Broken Arrow junior high school teacher who, with one of her classes, was doing a project on the 1921 Tulsa genocide. I remember being shocked when I first heard about this horror. I held a degree in history from Rice University and was a well-read person who had never heard of this horrible event. I cannot remember when, after I was hired, I first heard of this. After that, it increasingly dawned on me that I was mowing and irrigating grass and trees on the site of that disaster. Attacking white vigilantes had surged into the Greenwood community from the south and west overwhelming Black armed resistance and burning and killing as they came. What had once been the most prosperous and second-largest Black community in the nation was in one night and one day reduced to smoldering ashes.

Our 200-acre UCT campus, with parking lots and buildings, covered that exact central part of what had been Greenwood. The attackers also poured in from the west atop Sunset Hill. It was the high ground which dominated the community. Former Black soldiers from World War I fought and died trying the defend it. That ridge was now the northern boundary of the UCT property. When I first saw it in 1990,

Sunset Hill was completely overgrown except for the summit, which had been a clay quarry for the Acme Brick Company and was quite bare.

I began exploring this part of the campus. I only knew that it was within the boundaries of the campus. No one at the university ever told me anything about it or said that it was even something we were responsible for. The ridges terminal was precisely on the north edge of the UCT campus, and the view from there looked upon a very different Greenwood from what had been destroyed, then "renewed." I gathered my two-man crew, Gerry Wayne Ellis and Jerry L. Jackson, and the three of us explored the ten acres or so of this overgrown place, Sunset Hill.

In the very heart of a city of 400,000 with a close view of the beautiful downtown skyscrapers, we picked our way across the slope that was overgrown with Johnson grass, sumac, and greenbrier, seeking to cloak and bury its dark history. There was an eerie silence up there, even though we heard the buzz of the freeway and saw the hundreds of students parking their cars and streaming into their classrooms. The feeling was profound, like being in a different world. As I just mentioned, the hilltop had been the quarry of the Acme Brick company of Tulsa. I am sure many of its bricks were used in the rebuilding after the disaster.

When I first hiked up there, I found that the hilltop was nearly bald, reasonably level, and that our tractor could work it. There was erosion, but we were able to help slow that by building a few check dams. Vegetation was needed, so we planted the five or so acres of the hilltop in early December with 500 pounds of grain rye. We sowed the top with our big-wheeled spreader; then we drove the tractor, dragging weighted, metal mattress springs over the seed for leveling and covering. We had beautiful results and feed for birds and mice that next May. After seeing the emerald green results, I never returned to the former quarry. We still had to focus on our central landscaped areas, and for a year, I had little idea what else was on that south facing hillside besides the weeds, burrs, sumac, and briars.

When we did begin clearing, we found a grassed terrace of about four acres. The terraces interfaced on the slope with steeper parts of trees and thicket. The base of the main terrace was strongly built with earth and boulders about three feet high. And though the ground of the terrace still sloped down, it was smooth on top, and once cleared and mowed, was easy to walk on. I have no way of knowing, but I believe the terraces predated the massacre.

One of us found out that these terraces had been the site of three elegant homes, and when we first started to clear the land, we found three hand-dug wells and the outline of a swimming pool. These had been filled in by the urban renewers, but the wells had settled enough to stand out clearly.

Exactly where the houses stood, we were never sure, but it is certain that they were built before the fires of 1921 because no homes in this part of town were left standing after that disgrace. Soon after we began mowing, the beauty and humanity of the hillside opened like a flower. The houses were gone, but some of the plantings made by these three black families remained. Our first mowing was late winter, and along a small berm that must have defined a bed back then, curved a substantial planting of diminutive crocuses in full bloom.

I had never grown them, hardly ever seen them. Crocuses usually grow in refined places, carefully tended flower beds. They are medicinal as well as beautiful. Here they were, a very small, but striking planting, lost in a huge and sheltered place. When I saw them, they had survived for more than 25 years with no care, overgrown every one of those years by the Johnson grass and sumac. Crocuses are an ephemeral plant; they along with their relatives, trilliums, bloodroot, and golden seal, show up and bloom in early spring and then retreat to their underground parts as the world grows up around and over them. Hence their ability to return year after year.

Behind the crocuses, in a kind of glade, the west end of the terrace was framed with an ornamental crabapple. It soon bloomed with a deep reddish pink-colored flower, striking for that species and in this historic setting! I remember it as having good form as well. How I regret not having a camera in those days.

To the left, or south, of the crabapple lofted a 72-foot-tall pecan tree which grew on the edge of the terrace. Turning around, looking now east was a huge, unkempt crepe myrtle. It was 15 feet tall and 8 feet across. It bloomed and was striking, but its beauty, awkward now, because this species needs pruning to remain ornamental. Still, it was another survivor from the planting hands of a resident who gardened in the ashes. Wealth did regrow out of the Greenwood ashes, just less than there should have been. This terrace was a beautiful expression of that.

Turning around and gazing and walking east up the now mown lawn, the terrace narrowed and split. Uphill and north of our terrace was, to me, just big thicket with individual trees to fifty feet. Native tree species, which I had never studied and did not know well at all, grew there. These were outside of the manicured terrace and were thick with thorny greenbrier growing right to the top of the canopy.

Mock orange from author's house in Tahlequah, OK.

Too bad I was so inexperienced at that time, or else I would describe more to you now. Seventy-five feet east of the crepe myrtle were twin pear trees maybe ten feet apart and 45 feet tall. They made themselves quite noticeable by dropping hard sweet pears at our feet. I probably knew more than I am telling, but with books, doing, and seeing, I quickly became a better naturalist and nurseryman.

Anyhow, twenty-five feet east and up on the higher terrace was a spring bloomer, unknown to me then. Man, it was a great shrub, 12 feet tall, growing a little open amongst sumac or whatever. It had small attractive leaves and did not appear ugly in winter, but its flowers which opened after the crabapples were yellow-centered, white beauties with wow fragrance, and the many good-sized clusters lasted about three weeks; then the petals began falling and coating the floor of the shrub with an enduring white carpet. It was mock orange, of course. It's not a native plant at all and must have been a treasured hand-me-down, but from an improved cultivar prior to my uncovering it in 1992.

The twin trees that had dropped the good-sized, hard pears at our feet were to me the big find. You never saw pears like these in the supermarkets. When we took bites of hard crisp fruits we found them not to be astringent at all, but sweet and good, a little gritty. The following summer as I watched the fruit develop it passed through a stage where the little pears looked like apples, so much so that I had to remind myself that we had eaten the fruit last year and that they were indeed pears. I had figured out it was a kind of Asian pear, and then I eventually found it referenced as the apple-pear. I later found the same tree growing in Tahlequah. I'm pretty sure it is not grown by any of the big nurseries, not that there were big nurseries in 1930.

After the surprise of finding these historical treasures, we began caring for them. At the same time since the land had been urban renewed and left alone for 25 years, it was like a canvas for a landscape architecture. We put in an irrigation system and began planting. Some of the trees we planted on the site came from Mr. Stan LeMaster, a former Oklahoman, retired in Kentucky, who propagated historic trees and provided them to parks and public grounds across the country. Trees that we planted on Sunset Hill included a Jenny Lind tulip poplar, a Ming dynasty cypress, a Johnny Appleseed apple tree, and a Treaty of Versailles horse chestnut. From local sources I added a Creek Nation Trail of Tears peach and a Land Run black willow. Stan flew in from Kentucky, and he was very excited that our collection could become his signature achievement. Later, we increasingly felt touched by the sense that these were sacred grounds. There was a newspaper article. President Randle became interested.

In the Spring of 1993, after planning with my teacher, Rodger and Nancy, the nearby Emerson elementary school, the Greenwood Cultural Center, and others, we held an event and called it the Reconciliation Ceremony. I was the Master of Ceremonies for the day. One hundred and fifty fifth- and sixth-graders walked over from nearby Emerson Elementary, tying black-and-white ribbons together around our trees to honor the many killed along the very street they were walking. We gathered at our main front entry. There were two bands, one from nearby Dunbar Hunior High, the other the junior high jazz band from my teacher's Union School. Both sounded good as our entry was like a band shell. President Randle and lots of civic leaders were there. An official with the State Department of Agriculture presented us with a $10,000 check for a tree planting grant I had written.

Much of that I had slated for the hillside and the Greenwood Cultural Center. The featured speaker, a Ms. Lenora Davis, survived the genocide when she was 17, now at 89 telling her story to students, civic leaders and the press. Prominent local leader Homer Johnson reminded us of the damage done to the community by the 1965 urban renewal. What a day. President Randle came to the microphone and honored me with an embossed and nicely dressed citation, announcing that day as Sandy Mueller Day in our city, signed and proclaimed by our new mayor. After that our department gave out 400 one-gallon trees we had grown ourselves for anyone to plant as they liked; then we dug in our large post sign at the entry to the terrace "UCT Greenwood Park."

All this and a couple of nice newspaper articles, but no television. Back then I was disappointed and surprised; now I know something was up. I'm pretty sure that some people did not want our university to be associated with the 1921 genocide or the 1965 urban renewal.

I wonder when the city fathers began thinking about their university. In any case, they kept the land for themselves until our campus was first built in 1988. It stood right in the middle of Greenwood and controlled both Greenwood and Haskell, streets to the south and east of our hill. The Black community was getting nothing but the custodial and groundskeepers' jobs.

After our grounds department was eliminated in 1996, our work was given over to contractors. They removed a brilliant little pool and watercourse that fed the pond and an elegant long bed of daylilies and irises. This made their mowing easier, though they did add back a huge modern statue which they plopped right on top of where the Jackson Magnolia had been. But they did their worst damage to Sunset Hill, the UCT Greenwood Park. Our sign was removed, and the hillside was allowed for a second time to overgrow as if we had never been there. A huge new science and engineering building erected since I left is named the Helmerich Building, and the greenhouse on the Mueller Building, long since demolished.

We began planting after we put in an irrigation system for the entire terrace plus drip line to establish trees. This azalea bed would have become beautiful. It was well planted. The little tree to the right was a clone from a yet living tree planted by Johnny Appleseed, a gift from Stan Lemaster's historic tree propagation society to UCT's historic tree collection. It died shortly after I left along with most of our newly planted shrubs and trees.

We put a plaque next to our homemade arbor and couch.

Columbine in my setting reappeared year after year, always welcome, never invasive.

25

Five Years between UCT and Texas

My situation after I was fired was not good. With a feeling of bitterness toward the Teachers Insurance and Annuity Association and College Retirement Equities Fund, who handled the state retirement money, I took out the entire $10,000 of my retirement and vacation. Marsha and I spent a couple of thousand redoing the interior of her little house on the Canadian River. I call it her house because she had found it and bought it in June, 1995, eleven months before my firing. Of course, I had never been able to pay off my $4,500 loan from thirteen years previous, and I did not clear that lien until 2005. So my commute went from 64 to 99 miles, and I drove that distance every workday for my last eleven months in Tulsa.

I drove a Geo Metro and got 50 mpg; trouble was in the summer of 1996, it was just about worn out. Despite my connections, I found no good work in Tulsa. I could hardly live in a more remote spot for finding work. Marsha and I were drinking and fighting; I did porn when I thought it was safe. Arena was 21, and had her own place in Muskogee, but Graham was 15, Heath 13, and Rachael 12. The house was tiny but charming and on a beautiful bluff overlooking a silted up and meandering Canadian River. I met a younger man Eugene, who was a friend of Arena's with a big, heavy, old truck and a trailer. We partnered, doing private landscaping work in Muskogee and lifted some very heavy moss and lichen covered boulders onto his trailer. They were the key to our beautiful landscaping along with azalea beds, something I now knew how to do well. I maybe grossed $5,000, probably less. I got paid about $2,500 for irrigation jobs. Muskogee ain't that rich. I was, though, now treading water with a life vest on the very lake where I formerly drove the fastest ski boat. I had just been re-elected chairman of the Forestry Council, the one with the grants. So I took my plan to the Muskogee Public Schools and the Muskogee Park Department, wrote it up and got a $10,000 grant, the highest offered; it paid me $18.50 an hour to plant tree farms on six different Muskogee school yards.

I was assisted by the Parks Department, even with a bit of manpower and equipment. They were the official applicants, and I was their contractor. It was a unique and fancy 12-hour-a-week job, and it required a lot of driving and work on my part. It turned out to be a very beautiful program and fortunately was easily funded for a second year.

With this funding I lifted my income to about one-third of what it had been. Nancy Feldman paid me $500 to drive up to Tulsa to work on her place. I had a home, and Marsha made good money. I suspect that my age 50, combined with Type A, full-bull manhood running wild, made this time even worse for me. I barely remember family life or my children from that period. With Marsha no longer able to give of herself, I'm sure I was looking for girlfriends, but men in my situation, no money, don't get girlfriends. Thank goodness, actually.

As part of the project I worked with six different schools, the high school, the junior high, and four elementary schools. At the high school, I lectured the horticulture class about trees and the history of agriculture. The fifteen of them were able to dig the holes into which we planted 150 bare-root seedlings. At the other schools, I gave a couple of lessons about trees to two or three classes of third-, fourth- or fifth-graders, working with about 75 students at each school. I dug the planting holes, gladly earning $18.50 an hour doing so.

In the spring, several things happened at once. I was planting a tree farm at Creek Elementary, holes all ready, about 450 of them, I think. Three classes were planting, and they came out one at a time. One student placed the seedling, one sprinkled Greensand powder, and the third shoveled or pushed in the dirt to cover the seedling.

At that moment I ran into a camera crew who had been filming something else there. It turned out to be Tulsa Channel 6 reporter, Scott Thompson, filming a news show called Oklahoma Traveler, for which he was well known. Many of his shows featured trees and environmental efforts. I had actually been on his show about four years previously at UCT. I recall that show being on our historic trees. I am sure Scott remembered me that day at Creek. So he promptly set up and filmed the student tree planting, interviewing teachers, the principal, and myself.

A month later, timed to run on Earth Day 1997, we were his feature. I never saw it. Everyone at the Parks Department said it was incredible; it ran five minutes, unheard of. The biggest conference that I attended in these public years was "Keep Oklahoma Beautiful," held annually in Oklahoma City. They gave out awards for which there was much competition. Scott's segment won first in the media category and our Muskogee Tree Farm project won first in the education category.

Apart for the fame and glory of winning, there were no cash prizes and no other granting institutions, including the SBA, were interested in funding my format on a larger scale. My grant continued after the tree planting and through the summer with me watering and occasionally mowing the tree farms. That fit within my available hours of pay, and the little trees grew well, their roots restrained to the holes I had dug. The same children mulched the trees when they returned in the fall. Winter was wet that year, but not flooding, and in earliest spring,

February, after just the one growing season, I dug the trees whose sodden heavy rootballs held healthy roots and supported trees four feet tall with a trunk the width of your thumb, more or less. We had quite a few different species at each site. I recall redbud, bald cypress, Nuttall's oak, loblolly pine and lacebark elm among them.

I shoveled each sodden tree ball into a doubled-up Walmart bag and carried or drug the bagged dormant trees to the school sidewalk where I gathered them into their species and wrote that with chalk on the sidewalk. Later that day students came with their mothers and chose a tree to plant in their own yard. Over a thousand were taken; some ended up in public plantings. I used about 50 trees at a public housing site. All the trees I planted set in well. We felt that the children who participated and planted would make sure their tree was watered. At each site I left trees in place to provide much needed shade for the schools, actually for the children at recess. The year, 1998, turned out to be very dry in Oklahoma. I planted and observed our trees, and they did well though a favored Nuttall's oak planted on Corps land near my home died just before the first fall moisture. I did not water it, but required it to make it on its own. I think most years it would have easily made. Anyway, I know the fourth-graders watered theirs because years later I saw trees from the project scattered around town.

Apart from the success of my grant, life for me became worse and worse. My role in the family seemed diminished with my income. I still drove on family trips and participated in meals and house cleaning. I was mad for sex and got none, except for the early internet porn which left all sorts of trails for the innocent users of our computer to follow to everyone's discomfort.

In the summer of 1997, Graham, without my knowing much about it applied for admission to the Oklahoma School of Science and Mathematics in Oklahoma City. His ACT score was 27 or 28, or maybe he took it twice. He had excellent grades and recommendations from his teachers. By the time I knew more or less what OSSM was, he was in. Founded by the Oklahoma legislature, OSSM is the brain child of a prominent Houston educator, Edna "Iron Lady" Manning. The school, at that time selected 75 applicants from around the state to board at dormitories and walk from breakfast to their classes at a beautifully renovated old high school. They took seven full classes each semester, doubling the usual classes in math and science. They got good physical exercise, but there was no organized sport. All students began as juniors and studied calculus from day one. Most of their teachers held PhD's and were charming to us on parent visiting days. We sat in on a fake class or two, and from then on saw a whole lot less of Graham, except in summertime, when we often took trips to Chicago.

One highlighted an important experience of mine as a teenager among the rich and famous. It began after we had bought rag magazines and coffee for our drive north. It was midmorning, and I had left the driver's seat and was in the back with the girls. Our car was an older model Ford Station wagon, still pretty with its fake paneling and big enough for our six. I opened a copy of Star magazine to a full-page feature of "The Girl Who Broke George Bush's Heart."

In a quarter-page color photo, there she was, Cathy Wolfman, walking two manicured lap dogs near her $5 million Palo Alto home. She was fifty now, like George and I, and she still looked great. Cathy was in my class at St. John's. She was very good looking and popular. I did not know her that well or have many classes with her, but in tenth grade, I did ask her out on a date. To my surprise she agreed to go out with me, no doubt anxious to ride in my Studebaker. To my disappointment the Friday afternoon before our weekend date, she came and told me that an aunt had come in from out of town and that family stuff was forcing her to break our date. Oh well. Then two years later I asked her out again. Again she accepted and then, no suspense, she broke the date with the exact same excuse.

According to the article Cathy and George were engaged, rings and all, when she broke it off. The writer implied that George's down period of cocaine use and despair resulted from the breakup. No one was saying that it might have been George's drinking/drugging that caused her to drop him. Cathy definitely made the right decision. I know because I met George Bush. I spent an afternoon with him and two mutual friends. We were all seniors and our spring break was just starting after Friday school. Two of my rich classmates and I wanted to play tennis and needed a fourth for doubles. We were stumped until one of my buds said, "Hey, George is home on his spring break." We drove to some newer wealthy part of town and entered the kitchen of George H.W. Bush's Houston home. I met his mother, maybe she wanted to check out her son's local friends, or maybe she was just visiting with us while George dressed for our match. From the introduction, I found George to be arrogant and cold, at the same time somehow dull. I was the better tennis player. Young George never did seem to change much over the years. When we read the Star article, George was just Texas governor. Later when he became president, I thought, "Well, Cathy embarrassed me, but at least she didn't break my heart."

In late 1999 after drinking a gallon of cranberry juice and taking vitamin B to get some yellow back into the urine, I was hired by the McAlester hospital as groundskeeper for ten bucks an hour. I had an employee under me and some part-time summer help budgeted for me. I did my usual good work for them and was well regarded and for the most part popular. I was fifty-four years old entering the lowest period of my life and the low-water mark of my most debased actions. My new job had a 27-mile commute, long enough, but I was used to that. The hospital actually had substantial grounds that included a connected walking park of several acres. It wasn't as large as UCT, but similar in several ways. After a year I received a substantial raise to $11.50 an hour; no one thought I would get.

At lunch my first day on the job in the center of the crowded cafeteria, I saw a beautiful nurse dining alone and so I joined her. She was in her mid-thirties and carried herself with aloofness. Conversation was a little strained since I, a complete stranger, had dropped in on her table. It was crowded then, but I am sure she knew there were other seats. I managed to let her know that I was the new groundskeeper and found out that she was the charge nurse on one of the floors.

My obsession began as a tiny figment in my sex-addled mind. After that I ran across her in the break room for smokers and once she marched in to our facilities department tool room because she needed something fixed up on her floor. She had all the men wrapped around her finger. She was divorced and had one child, a son who was 18. It turned out that she wanted me to hire her son for a summer position that was in my budget. Even this normal and innocent proposal on her part fed my dreams of connection with this woman. I did hire her son, although my boss had employed him the previous summer and suggested he was not much of an employee. The pull of this obsession now got a lot stronger. I became a father figure for the lad, and he did decent work for me. Now I was more involved with her through his stories and trips with him to their house. I soon picked up an outside landscaping job for one of the physicians and needed my young man for a helper.

One day I used my connection to offer to come over to their house and help them work in and plant their garden with some flowers I had. It was almost like a date; she came out and worked with me in the yard. We chatted. She never dated anyone from work. I had a miserable marriage. I did not tell her I was sexually frantic and all my cravings were flowing through my imagination, an imagination further fueled by porn into her psyche. She was a magnet for my attention and imagination.

At one point, I apologized to her for being too personal. Too bad I didn't keep the promise. I tried to impose my will on her, making comments on her life. We even had a conversation once about "will," as if we were two magicians, one trying to control two, two trying to make one disappear. I was reduced to a mindset of rudeness and passive aggression. In the midst of an ice storm, exhausted from overtime, overloaded with pot, I wrote her a note, one that incriminated me. I was promptly fired and evaluated for any possible danger I might pose to her. She got a restraining order for herself and her son, and I paid the $115 court costs.

In retrospect this woman was very similar to my own mother. There was a strong physical resemblance in face, form, and demeanor. On one of the court papers, my nurse's birthday was typed, July 18.

That was my mother's birthday! Were the stars setting me up? How low could I go? How could I possibly behave in this fashion? Who was I?

Shocks are good for self-observation and awareness. About this time, I was reading or rereading *In Search of the Miraculous* by P.D. Ouspensky. That was my first connection with the Wisdom of the Fourth Way. It asked me to "know myself." The shocks kept on coming, of course. I felt disgraced in front of my children and with Marsha, humiliation. I didn't like to go to McAlester at that time for fear of being seen by a former colleague. The more I came to "know myself," the more disappointed I became. While working at the hospital, I had rented a home nearby for $200 a month to separate myself from Marsha and her tiny home. Now, I had to beg for readmission to her home. She took me back in, even disgraced. I took a job just four miles away at Narconon, $6.50 an hour to push a small mower all day long over endless turf at their giant facility.

It was good exercise for a 55-year-old. I got fit and tanned. I worked there till the end of mowing season when they said they were releasing two of the four of us. I jumped at the chance and resigned immediately. They did not like that. They had wanted to keep me and see the others go. Then I applied for unemployment benefits and they had to pay me. I needed it.

One morning I was alone working outside at Marsha's house when I heard a very loud and strange watery sound down at the river. I hurried down the thickly wooded 200-foot slope to an outlook above the Canadian River. I was looking down on a channel of the river about 30 feet below me and no more than 50 feet away. The channel was about twenty feet wide at that point and the far side was a sand bank not yet grown over with cockleburs. There were about a hundred white pelicans. Most were on the sandbar, but about twenty were in the water. I watched as they formed themselves wing to wing across the channel and then the entire crowd began swimming, slowly moving up stream and splashing as they arced across the channel at the same time. They were driving stunned fish into the bank and feeding. It did not take long, five minutes of frenzy and then the diners walked on to the shore with the others. Then one of them ran a few steps and lofted into a lazy spiral. It was followed closely all the other pelicans. It reminded me of humans boarding an airplane. They looked like a kind of giant bee swarm, and I bet they were three or four hundred feet up. Then they sort of gathered into tight formation like a mob and dive-bombed another channel about a half-mile upstream. I could not see the strike, but I sure heard it, that same unique sound.

Nine-eleven happened. I watched it over and over, fascinated, horrified of course, but deeply cynical, and I perceived the deep-state minions and the Rockefelller/Rothschilds as the perpetrators. My sister wrote or called. She was doing well in Austin and finishing a large home on five acres in the foothills above Barton Creek west of town. Our brother was the carpenter/contractor. The two of them and my mother were renting a large apartment in town with room for me. Could I come down and work for my brother as handyman/helper, room and board provided?

I arrived in Texas in October, and a time of great healing began for me. Over the next six years, Marsha and I divorced. I held two good jobs and found welcome female love. In the end, both jobs ended in disaster and the love was fleeting. But there was a Fourth Way group in Houston, and my study of Who am I with them helped me see the bittersweet view of my personality in action. I came to identify the state of waking sleep when my personality, that thing I learned growing up, not me, takes over, including the foot fetish. The more I saw it, the more capable I became of letting it fall away, stilling the body, finding attention in my mind and focusing it on sensing the relaxed body, the breath. Doing that I am closer to God and my own "still four-year-old" essence.

A solid wall of bitter melon. You can just make out the clear bright yellow flowers. When the state inspector came, he marveled over the beauty of this plant; he had never seen it before.

Up close, the fruit. Seed is coated in a red pulp which is sweet and good. Too bad the fruit is inedible to all but a few aficionados. I enjoyed a tea made of all parts of the plant.

26

Texas, Our Texas

It was such an easy arrival. My mother, my sister, and my brother were very glad to see me and have me stay with them.

Sand's mother

Mostly I helped my brother John move ladders, caulk, paint and sweep up as he finished construction. He wanted to pay me, but by declining that, I retained my ability to do other things, like look for work. Besides, my dear brother John was too much of a perfectionist for me. By Christmas, we were moved into Ann's new house. It lay on five acres which drained into nearby Barton Creek and had three levels. John occupied a large comfortable half-basement with an attached and very convenient garage. My sister occupied an elegant bambooed upper floor with patio and good views of the Hill Country. There were two, more than adequate, bedrooms for Mother and I on the main floor. It was the most elegant and comfortable home I ever lived in, and later after I moved to Houston, there were many weekend visits to see the three Muellers.

From there I found a part-time, $10-an-hour job at an ugly little wholesale nursery hidden in the live oaks on a slope that ran down to the same Barton Creek about four miles downstream from my sister's. That job quickly became full-time, and my Spanish was a big plus. I am guessing that all ten of his lady planters were undocumented. By Christmas I had this job, an older lady friend for a companion, and the comfort of my mother's cooking. I also attended The Church of Conscious Harmony, which had spectacular architecture, a great bookstore, Sufi dancing, and a study of Georges Gurdjieff.

I had no intention of staying at the dirty little Barton Creek nursery, so I began looking at the ads. I was offered managerial employment for $35,000 a year with Perfect Lawns of Austin. Everything was done chemically, each spraying was a sale, as was every plant or prune. And while you did do the grounds work, the real tzismus of the job was to be on the phone in the evening selling all the prunes, plants, sprays and remodels. I drove around with them on the day route and saw the little offices with phone and computer. Met the Design Department. I could never do this job; I almost trembled with relief that I didn't have to take it.

At the same time, I knew I was leaving the Barton Creek nursery I did a criminal act and fortunately got away with it. In a crummy shed where the nursery kept its pesticides, I found a 25-pound bag of oxamyl. This is an iso-cyanate compound of the type made in Bhopal. This formulation is put into the soil to kill eelworms and soil pests. It has no legitimate or registered use for potted plants, so I do not even know why it was there. Oxamyl is particularly dangerous when it enters a riparian environment as it is devastating to aquatic life. Warnings on the bag specifically refer to this.

The nursery sat on a slope just 100 feet from Barton Creek, which is Austin's most famous creek. It flows through a beautiful canyon and feeds the famous Barton Springs swimming hole before it empties into the Colorado River near downtown Austin. Tens of thousands of Texans bathe in and enjoy this waterway. A single strong storm could have washed this entire bag into the creek. So I stole it. I stole it and took it to a hazardous waste disposal site and paid $70 to have it placed into a sealed container and buried in the safest way possible. I am proud of what I did, but also certain that all over the land hideous chemicals end up being carelessly allowed into the environment because people like money and convenience more than they like God.

Before I stole the poison and left the dirty nursery, I found two advertised positions that turned out not to be in Austin, but north of Houston. One was offered by a man known as the King, who took his valuable Piney Woods and turned it into the world's largest Renaissance Faire, open just four weekends a year. The King was well known to thousands of young people from the Houston environs who had become the ale house maids, blacksmiths, or gypsy souvenir vendors of the Faire. They were also the princes, princesses and jousting knights. Most of those people, according to the internet of 2002, did not like the King. When I interviewed, I found him to be strange and the job too high-tech chemical for my liking, so I told him I wanted $50,000 to work for him. He said he was sure I was worth it.

I had also arranged to meet the owner of a Catholic retreat center called Circle Lake, who pending a background check hired me on the spot. Ralph Marek, the youngest of three brothers, was seventy-six years old in 2002 when I met him. He had grown up in the rural depression of East Texas and was familiar with eating squirrel and armadillo. The family fortune began when the oldest brother started a carpentry/contracting business just before the outbreak of World War II. After leaving the armed forces in 1945, the brothers used their tiny start, and a new industrial product made with gypsum and paper called drywall, to become the largest interior contractor in Houston and probably one of the very largest in the country. Someone said that the Marek Brothers built the insides of every downtown skyscraper and every large commercial building in Harris County. And the men who did the work were Mexicans. When I started, Marek Brothers had expanded to include a massive employment agency supplying temporary and emergency help to all the other contractors. The Mexican men did carpentry and landscaping, and the women were maids and cooks. Greater Houston probably had a million residents when I left in 1968. Now there were five million. Ralph was just then building a new cathedral for what was now to be the Archdiocese of Galveston-Houston. It was plenty strange to me to be back in Houston after a thirty-four-year absence.

I started in January 2002, and from the onset I had two bosses; at first, mainly Ralph, but at the end mainly the Archdiocese.

Circle Lake was in Montgomery County north of Tomball. It was 45 acres of sandy soiled, piney woods looking down on Houston from an elevation of 190 feet. It was a botanical garden without the signs and was originally the weekend getaway place for the three families Marek. There were three large older homes which housed each branch of the family, all with a couple of bunkbeds to house the many guests. Those were on the east side of a circular lake which created a near island and named the property. The lake was good sized, about four acres, and 12 feet deep in parts. It was popular with fishermen who paid five bucks for the privilege. The banks also housed a colony of nutria who also fished and damaged landscaping with their tunnels. We hunted the nutria with .22-caliber rifles when guests were not present.

The east side of the property was bordered by the main railroad line coming north out of Houston. It was thickly landscaped with camellias, azaleas, maples, shrubs and bulbs of all kinds under pine and a few large magnolias. It had an old but usable irrigation system which pumped from the lake. The larger west side of the property was open except for a few large trees and a row of massive crepe myrtles which lined the entry drive. This drive landed at the office which was a modern, modest, three-bedroom house, also the home of Sister, who registered the guests and collected their money. North of this house was a small parking lot and a cafeteria that held up to a hundred people. Down the drive was a second larger and nicer home on a pond at the secluded low southeast corner of the property. When I arrived, Ralph's friend Larry and his wife lived there, and he was the onsite property manager. Larry and Ralph had interviewed me together when I was hired.

Also employed at Circle Lake were two tiny men from El Salvador, Manuel and Leo. Ralph was worried that Manuel would react badly to my employment and the negative effect that might have on his authority. Manuel did, in fact, turn out to be a Gallo, and we were not close but never had any serious disagreements or public spats. Manuel's helper Leo, a rotund, yet fit young man, lived on the property with his wife. My apartment was about 200 feet from Leo's place and we became good friends. Neither of us had decent accommodations. Leo and wife lived in a tiny trailer which shared a wall with a shed full of old equipment and mice. I fully earned Leo's respect one afternoon when he came to me very upset and describing a large dead snake in the shed and an unbearable stench in his trailer. It turned out to be a 6-foot rat snake that had gotten trapped in plastic netting, and I ended up cutting it up and pulling it away from its bane. This was made much harder by the rotting condition of the corpse.

My apartment was a moldy, three-room affair, the corner of one of the original three family homes. Larry lived a good five minutes by golf cart away from our dwellings. Nevertheless, he soon became aware that I smoked marijuana and after an elaborate sting operation involving a fellow visiting me who heard and retrieved the ringing of a phone somehow lost in a bed outside my house – ah, a lost phone – which my visitor set on the counter and then began to talk about procuring large quantities of pot. At that point, I accidentally tapped the phone into my sink full of sudsy dishwater. I got hauled before Ralph anyway even though I had destroyed the evidence. I did confess to smoking. He had the good sense to probation me, with urine testing at his manpower business. He also gave me an inspirational book that he favored, and his counsel to read it. I did, some. Ralph was a sincere and caring man, in a Roman Catholic sort of way. I did like him, and he liked me as well, or certainly the good work I did for him. He let me know on several occasions and sometimes with temper that he was boss. The Catholics paid my salary of about $34,000. Ralph gave me, from one of his accounts or businesses, an additional $300 a month cash. That had the effect of keeping me attentive to and supportive of his Saturday extravaganzas.

Circle Lake, being both religious and nonprofit, was eligible for and a recipient of court-mandated community service from the offenders. We had a substantial number, around a hundred who owed the community anywhere from 16 to 1,200 hours. They came in on their own during our open hours, and we put them to work and on the clock. During the week it was rare to have more than three or four show up, but on Saturdays there might be thirty. Ralph always put two or three women to work in one of his kitchens to prepare lunch for all of us from the canned, donated or reduced food that always seemed to be flowing out of his cornucopia. Ralph arrived early, usually bringing one of the Mexicans who worked with him at Marek Brothers. He had gone either to Lowe's or to his son Tommy's nursery and filled his truck and large trailer with plants. If the plants came from Lowe's, they were deeply discounted slow sellers and everyone suspected that Ralph got better than good deals on these plants. He loved bargains. He also loved flower beds everywhere and some of the stronger young offenders could always dig more and haul more. Some of those beds were so helter-skelter they became pretentious like Mr. Barnum and Mr. Bailey.

These Saturdays could become very hard to manage. I was the number one foreman. Ralph helped, but always napped after lunch when the day was most difficult. For a while I had help from our newly hired facilities manager, but the sewer plant made him ill and this important help was lost for much of the time I was there. Larry also helped at first, but he left soon after I came. Saturdays usually left me exhausted.

Coincident with all this on the open west side of the lake, Ralph was building a large and lovely conference center, adequate for about 200 people: thirteen three-bedroom houses with dormitory style bedding, which along with the multi-bedroomed, old homes offered bunks for space for more than 200. He also was building a lovely chapel on the island and an elegant-windowed kitchen and dining hall on the old east side of the lake. So Manuel and Leo continued with the maintenance of the established landscaping, and I was in charge of all the landscaping installation. As the houses were built, we closely followed with irrigation that encircled them and extended to the acres of continuous lawn. By this time, I was quite skilled in design and installation. Eventually, every house had four-foot wide flower beds extending on all four sides. We filled them all with shrubs, perennials and annuals. All the leftovers went into our 90-by-40-foot greenhouse.

It was an orgy of planting. Partially because of, partially in spite of, all the flowers, Circle Lake was a very beautiful place. It was a showcase of perennial bulbs and roots, decent lawns, beautiful understory, massive magnolias, redbuds, dogwoods and tall pine and cypress. The acres were mostly level but fell away toward the creek in the southeast corner. Amidst the beauty of this nature, one would often be reminded of the rail line because the big laden trains had built themselves up to 50 mph by the time they reached us on their heavy-duty track. I loved their roar and sometimes would race to the fence to watch them up close. Eventually the church reluctantly gave me one of their rentals, a cute one-bedroom cottage with amenities and no mold. The deck opened to a view of the tracks, filtered through vegetation, but only about seventy yards away from the roaring monsters. As part of the deal, I was not to have overnight lady visitors.

I remained madly hormoned throughout this period with no relief. My counselor and Fourth Way friend, Bob, suggested I go to Sunday services at Houston's elegant Unity Church and attend the singles brunch that followed. I did that, met Sandra and was with her for the next four years until her death in December 2006. She was two years older than I, still lovely. She told me at the beginning that her liver was damaged beyond repair by an antibiotic she had been given years previously. She was on the liver transplant list, and I went to classes with her for that hope. She functioned well when I first met her, but got worse over time. Her home was in Sharpstown, a development that was a big project in 1955. I remembered the neighborhood from my high school days. It was modern housing for the modest. Sharpstown had very nice little homes, with one-car garages and adequate fenced yards. It was almost immaculately clean. Only a few larger homes were 1,500 square feet; Sandra's was 1,100.

Now after fifty years of rain and sunshine the subdivision had a beautiful canopy of large trees interspersed with palms and flowering shrubs. In that very place, I saw a stupendous plant. It was an Agave Americana in full spike before the flowers opened. I first noticed the bloom spike on a bike ride, just a few houses down from Sandra's. It rose like a tree trunk over a neighbor's roof from their back yard. It was more than twenty feet tall! I peeked at it from their back-fence gate, and the plant itself was about eight feet tall and had suckered to a width of about 12 feet. The central plant would die as the seed ripened, but there were so many suckers I'm not sure one would notice. In a few more years, this clump would be throwing up five of these spikes at a time. It could take over the entire backyard or else make a lot of pulque.

Unfortunately, Sandra's house was a long way from mine at Circle Lake. Since she was willing to move, and I was already committed to her, I brought her up one quiet evening to spend the night at my abode. Somehow Sister saw her, and I came under the discipline of the church. Booted out of the cottage, I went to Sharpstown to live with Sandra and entered another period of long commuting.

Except for Saturday, Circle Lake was a very enjoyable place to work, and I could almost always define my tasks for the day everywhere surrounded by beauty. The community service workers were of every imaginable personality, race and background. Many enjoyed the physical work outdoors, others brought useful skills, since there were plenty of electricians and mechanics who got in trouble with the law. One was a wealthy car dealer with a big home near where I lived later. He had me over for barbecues and birthdays and paid me to do irrigation work. He had a big mouth and used it on the judge with the resulting sentence of 1,200 hours! Usually if an offender pitched in, we would credit them with more hours than they actually worked. We credited a lot of eight-hour Saturdays as twelve. I was glad that Ralph and Larry had established this policy. Some of our helpers had been given more than enough community service hours. As long as we documented well, I do not think the courts cared. Almost all the offenders manifested good intentions towards me and I reciprocated. They carried a kind of vulnerability that sometimes allowed them to express their most positive attributes. A few of them became personal friends and one was my car mechanic! It had been the same with the work release inmates at UCT. They were grateful for kindness.

For me and others, it all got ruined when the Church sent in new management. Larry, in spite of the marijuana issue became almost a friend. We certainly worked well together. But in the fall of 2002, he and his wife left their nice home and volunteer role at Circle Lake. Bob Schmidt and his wife, Velma, moved in and took over operations. The diocese was taking over more of the operation from Ralph, and he was fine with that and backing away somewhat as well. The big boss, Brother Jim, who assigned the Schmidt's to Circle Lake, was an up-and-comer at the Archdiocese headquarters in Houston. Jim later put on a conference for all diocesan employees. Perhaps four hundred of us spent a day and a half with good meals at a swank Catholic center on Buffalo Bayou. The bayou is a high-dollar place in Houston. It ran next door to my old high school, and I still have millionaire friends living in River Oaks, heart of the bayou beauty. The conference was fine, and at first, I thought Brother Jim was an honest and caring person. Bob, however, was a jerk from the beginning. He dismantled the cooperative and supportive offender program Ralph and Larry had established with anal retentive relish. He took over control of the record keeping with the auspice of a Nazi official perusing the papers of Jews. I was very surprised. How could anyone be this way?

Bob was of Polish descent, a midwestern devout Catholic who had found his way to Houston as a member of a group of that, while non-clergy, were connected to and employed by the diocese. He was about my build, a little heavier, but angular and gawky. Bob's demeanor was a little strange and people commented on that. They spoke of him as a phony. I worked hard at being conciliatory with Bob. Occasionally, we agreed on things or complimented one another, but those were respites in an atmosphere of hostility.

In one of our weekly meetings he gave us (Us at that time was Bob and Velma; myself; Jack, the new and short-term facilities manager; and Pat, a diocesan counselor who then lived in my old cottage), the Briggs-Meyer personality assessment. Bob and I were exact opposites and both of us extremes. I think I was ENFP (Extraverted, Intuitive, Feeling, and Prospecting), sort of like floating above the planet extreme ENFP. Bob was everything clerical and by the book, like he lived in his coffin.

Bob's wife, Velma, was of Mexican descent. She was a very big woman, tall and obese. Velma, like her husband, was employed by the diocese at Circle Lake. She was in charge of guest services, and while she did not often need offender assistance, when she did, the ladies I conscripted to her service complained of her demanding and unappreciative management style. Bob and Velma were to say the least an unlikely couple. Surprisingly, they had a daughter. She was nine years old and very genetically damaged with difficulty walking and speaking. No one who met them could possibly doubt who was in charge. Bob was a phony and very passive aggressive. Somehow, he never developed the confidence to relate to another person with sincerity. We considered him weak willed, but he clung to rules and regulations so strongly that he ultimately destroyed the employment of everyone who was at that meeting, including Velma and himself.

Ralph was unhelpful to me, though he had several spats with Bob over our Saturday work fests. He wasn't pulling his weight to stop "the Bob." At that moment, my relationship with Sandra offered an out that seemed good to both of us. Sandra wanted to move to Baton Rouge to be near her daughter, and Baton Rouge Schools were appealing to persons like myself to teach there. My employment was guaranteed if I passed two fancy tests which were numbered, monitored, and certified. One of those was math, and I was still somehow good at it. As I say I was still unhappy at Circle Lake; events moved rapidly and without a fare thee well. I left Circle Lake and the now-archdiocese. I did let Ralph know why and where.

I liked Baton Rouge and the Cajun culture, music and food. I liked Sandra's daughter, her husband and boys. I stayed in their big house while Sandra worked to sell hers. She came for a weekend, and we looked for homes. The one we found and would have bought was very substantial and appealing. All I had to do was become a successful teacher. Become a successful teacher in a shockingly bad school, that is.

Baton Rouge Public School children were almost all African-Americans. The white children went to an astonishing number of private schools, so in Baton Rouge there was de facto segregation. In most places private school teachers earn less money than their public school counterparts. That was certainly true for me in Jemez, but in Baton Rouge, you got a little more money to teach the white children. Naturally, the only job openings were with the public schools. My pay was $27,000, and the assignment was to a junior high school that had an AP math class that required my math endorsement, something no teacher had at that school. I would teach four other math classes, only two of which had the same prep. The school itself was a tight collection of three dirty red-brick edifices; the cafeteria was one story, low ceilinged, and noisy, and the classrooms were tall ceilinged with window glass up high and fans. It was definitely reminiscent of the Chicago schools I had worked in thirty-four years previously. The building was not a bad space for teaching at all, but it had a feeling of dirt and decay. I do remember that the Chicago schools I saw or worked at had great teachers and solid principals. That was not the case in Baton Rouge.

To me this school demanded a career-oriented Black principal. Ms. Moffat, my principal was white, about 50, and didn't seem to care much about anything except appearances. The district said that I was to have a mentoring teacher and supportive reviews from the beginning and that I was to teach the AP class. None of those things happened. The mentoring would have been great, but I was never told anything about that and no mentor ever showed up till the day before I quit. That's how I found out about it. As for the AP, that class had been stolen by a returning math teacher, not legally allowed to teach it. He, of course, wanted the interest and good behavior that comes with a class like this. No one mentioned this detail to me. In fact, I was unable to procure all the books I needed for this class, not knowing my thief teacher had already taken most of them for his class. I had three seventh-grade math classes, and the period one class seemed the brightest of those classes, so I unknowingly began teaching them the AP class. We shared the books or I did freelance.

I knew at the time that I could have done much better if I spent more time preparing and organizing. I pretty much had to do my own curriculum for four different classes; there was no departmental anything. I spent two or three weekends at the beginning driving to Houston Friday evenings and back Sundays. Houston-Baton Rouge was a very long drive. I remember I got too tired to prepare good lessons.

But my classes were by no means a complete failure. One chaotic class of eighth-graders held within it a sizable corps of bright young women. I assigned them algebraic expressions and problems that weren't normally taught till eleventh grade. The girls were smart and loved doing it. My "AP" class seemed to understand well enough, and my other seventh-graders were mostly doing their lessons. All my classrooms were too loud, but papers were turned in. I realize now that I spent a good deal of my prep time grading their papers in thin red ink, noting their errors and or working the problem for them to see again how the problem was solved or what a symbol meant. I often added words of praise or encouragement, addressing them by name on their returned graded paper. I encouraged them to take the papers home to their parents. I pinned their best papers on a big bulletin board which became a big hit with all the classes. I also began calling parents and was getting appreciation and help for them.

I found out finally from Ms. Moffatt, that all graded material had to be kept by the teacher like it was a legal document and at the same time advising me that fewer grades were better. Oh, dear, most of my graded papers had gone home to the parents, horrors! So much for my best work and ideas! Most classes I kept going for half the period anyway; weak lessons, insufficient practice material and my failure to provide a quiet routine for those who finished early meant uncomfortable noise and class movement. Nothing that bad ever happened, but Moffat, who could hear whatever she wanted over the intercom, finally burst into my class, was shocked that I had dimmed the lights, and where were this class' lesson plans? They should be posted on the door. She stood there glaring while the now seated students looked up at me, and I had them take a sheet of paper, fold it various ways, and mark fractions. Finally, she left.

At this same time, she announced that the first half of the nine weeks would not count, and in fact, whole new classes would be formed. Whereupon, she approached me, as if no hard feelings, and said she needed me to teach a chemistry class, did I feel good about the subject matter? I should have said no and gone to the fellow who hired me screaming bloody murder. But I took on the class with no certification for chemistry. It would just be the remaining four weeks after all. There were three days of horrible classes in chemistry in a big chemistry classroom, lots of students without enough of the crappy books. I literally could do nothing with them and had nothing to teach them from. Moffat came to see me. A parent of one of the chemistry students had complained about my teaching, and therefore, an official from Title 9 was going to interview me. Title 9 was a federal program, and they specifically were going to be talking about my lack of chemistry certification. I should have let that happen; the heat would have been on Moffatt. With a sense of relief, I handed her the classroom keys and gave her my verbal resignation. She was irritated that I wouldn't even work on long enough for her to get a sub.

My quitting was enabled by events at Circle Lake. Ralph had enquired by email if I knew of any sexual allegations against either Manuel or Leo. Eventually, it turned out that they were both fired for misconduct. Ralph was willing to push his weight and get me rehired. So after five weeks, I returned to Houston, disappointing Sandra, although we stayed together as before. Bob extracted his pound of flesh.

I was hired back on some temporary basis and paid less than what I had been making before, dropping from $34,000 to $27,000. I'm sure there was a write up on me in my personnel file. When I returned, I found that my friend, the oft-absent facilities manager, Big Jack, was gone amidst a lawsuit, with Jack claiming the Archdiocese had poisoned him by making him clean out our faulty septic plant.

In response to the sexual pranks of Manuel and Leo, Bob kept all women offenders off our crew, but after a while that went away. The worst thing Bob did from my point of view, though it made perfect sense under the circumstances, was that he hired a mowing crew. In removing the mowing from my responsibility, he gained more mastery over me. I was no longer critical from a landscaping point of view. And his work crew could also be hired to do things Bob's way or special projects that he liked, or to pitch in if I were gone. Ralph and I were left with the planting and watering. Bob was so distrustful of my giving extra hours to the offenders that he actually hired a former offender, Johnny, a young man with little education, someone who might be called a friendly simpleton, to maintain all the offender hours. Ultimately, that idea failed dramatically with another "hours for sex scandal" at the hands of Bob's offender employee, and after that I pretty much returned to that duty, usually being the only staffer working with the offenders.

Although Sandra and I remained paired until her death in December, 2006, I moved out of her home and rented an apartment much closer to Circle Lake. My daughter Rachael came and lived with me briefly there, but soon got an opportunity to return to Oklahoma. At the end of 2004, Sandra bought a lovely, big, 1,750-square-foot home in a northern suburb past the Woodlands. We moved in together, but after about three months, she took a turn for the worse and moved back to her unsold Sharpstown home. Her son worked at Methodist Hospital, which is near Sharpstown, and she needed the help and convenience of being close to help. That left me alone in the big house, paying more rent than I wanted to, but not wanting to leave her with the full mortgage payment. I stayed there almost a year. I had a couple of flings in this period, but continued visiting and helping Sandra with the shopping and chores on weekends.

In the meantime, with Ralph's subsidy, my salary rose to about $38,000, and I maintained a truce with Bob so that day-in, day-out work wasn't bad. We did a lot of big projects after the main development had been landscaped. We rehabilitated and drained some waterlogged beds and designed and built new ones. My irrigation work was spot on and so was the soil work. Ralph liked me more and more.

But suddenly I got written up. I remember it being some minor and unfair thing. I complained to Ralph, who sympathized, but told me to let it go. Then a few months later on a frantic Saturday with virtually no help aside from the resting Ralph, two offenders took a joyride in a new golf cart I had been specifically obligated to protect, and they smashed it up in the woods by the creek just off the property. It was a several-hundred-dollar repair, and I got my second write-up. It was late on Saturday, and I just left the wrecked cart where it was, planning on towing it out Monday. Bob, however, went down with somebody and brought it up Sunday. On Monday, he informed me that by leaving it in the woods, I allowed them to come back and steal the eight expensive batteries. I did not believe him; it was just too remote.

Then a month later, I went in to his garage for some company tools and there were the eight batteries. He actually perpetrated this scam on me to make my second write-up even heavier. After that Ralph worked on a way to remove Bob and Velma from our management. He managed to get Velma transferred to a woman's shelter he funded in the valley near Velma's family; that in turn left Bob dangling. He couldn't divorce. Ralph may have known that Bob and Brother Jim were more than friends. That made sense to those of us who knew them. Johnny, Bob's pet offender and record-keeper, showing no favor to his benefactor, told us he overheard a phone call with Brother Jim. He suggested that the conversation seemed personal and which Bob ended with the salutation, "Bye-bye, Sunshine."

I knew nothing but intuited it to be so. By October, Bob was readying himself to move out. We all felt relieved and hopeful for a better, more human, replacement. Sandra died at Christmas. Bob was gone helping Velma move. I needed just one day for her memorial and for the sprinkling of her ashes. Brother Jim had to be contacted because it was a day I was needed for work. He refused to give me the day and I told him that permission or not, she had been my partner for four years, and I was taking the day. That set the stage for Bob's return for his final weeks.

Just after New Year's, while working the tractor in a new garden, I lost my keys. Earlier I had asked for a key chain that looped securely onto my belt but ended up with a clip-on one. It came loose. I had no idea where the keys were, although, I heard from Ralph in the aftermath that my keys were found in that garden. I had had a key incident before when I couldn't find my keys at home, so I knew Bob was sensitive to this. He gave me till Monday to find the keys. When I came in without them, he gave me the third and fatal write-up. He advised me of this with a kind of smug glee. I could have appealed. Maybe Ralph could have saved my job. Instead, I treated Bob to a volcanic eruption in the presence of a young new secretary who had just started. I concluded my rant with a warning to Bob that someone would "cut him a new asshole," that it would not be me, but that someone could do that. He threatened to call the police of course, but I told him not to bother I was leaving. Once again, my anger was volcanic, probably bad judgment because Bob was leaving and I had had a chance to stay on.

On some level I was ready to leave, so I had prepared a landing spot. Ralph's younger son owned Magnolia Gardens Nursery. There were 150 acres and over 100 employees. I had gotten to know some of them over the years, and in a chat with Ralph's nephew, John Marek, manager, I was assured they would hire me if I left Circle Lake. So after a couple of weeks off, I began another long commute, this time to Waller, Texas, on Houston's northwest flank. The nursery actually was in a rural setting with some open fields and scattered housing. It was definitely west of the Piney Woods I. As usual, they took their pound of flesh; I started at $30,000 and was assigned an ugly task from the get-go. John was unhappy with the current perennials and propagation manager, Patricia. They did not want to fire her and hire me for what was the No. 2 position at the nursery; rather they wanted me to impress them with my work and dig up dirt on the embattled manager at the same time. I began at the exact center of the problem, with Pancho and Antonio. Antonio was the waterer for the entire twelve acres of perennials and one of the three large greenhouses. We became like brothers because I understood enough of his job to see that his work could be done no better given the system we had. Often nursery mangers come to see their long-term waterers as untouchable, critical employees. That is true only when the manager does not know the watering himself and can't step in when needed. I think there are managers who are either too weak or too "important" to know the details. Anyway, it was bad for Patricia that Antonio didn't like her.

Pancho was from a big family with farm acreage near Guanajuato. He was good sized, shorter than I, but much stronger. He was in the prime of his manhood and was the respected leader of the entire Mexican contingent working at the nursery, about 75 people. By Mexican, I mean all the workers who worked about nine months a year with an H-2 visa. There were a few Texans of Mexican heritage who worked all year, but not many. No "illegal aliens" worked at Magnolia Gardens. Pancho was in charge of the critical job of taking, sticking, and rooting cuttings, usually working with a team of four ladies. Their production for me later was outstanding. Patricia and Pancho had had many disputes about how to do things. Their dispute was worsened by the fact that she had far less understanding of propagation than he had, yet insisted that things be done her way. Before I arrived, she had demoted Pancho from his assistant title, giving that role to Roger, a legal Mexican-American whose Spanish was not nearly as good as mine. Roger was nowhere near as competent as Pancho, so it became a kind of charade with a share of antagonism.

I often think of Pancho to this day. We were closer than brothers. My Spanish made a huge difference, but it was more than that. We saw the nursery world the same way and loved growing plants. From the beginning I protected Pancho from Patricia and gave him hope.

Patricia had gotten her fertilization injection rates way too low, by a factor of about ten, so that was brought up by me as well, and I also pointed out some really excessive pruning she had ordered. I don't know how obnoxious I was; it can be ugly to point out other's mistakes to their bosses. Eventually, she went to John and told him either I left or she did. And on that day, I became de facto propagation manager, number two man on the nursery ground. There were eight acres of greenhouses, three for propagation and five for one-gallon perennial production. There were also four acres of outside beds for hardier perennials. About 25 people worked under me. Of course, John never offered me the actual title, and I had to insist on my pay being brought up to an appropriate level, $38,000.

From the beginning I had great success. Pancho overruled me and kept the propagation house hotter than I liked. He achieved very high rooting percentages even with difficult cultivars. I had my role also, researching and mixing rooting solutions appropriate for each species of cuttings. Pancho and I worked the timers together, and Antonio worked inside with a hose when edges started to dry out as the cuttings rooted. I kept the fertilizer flowing throughout my domain and personally watered the growing plants in the main production house, all on a timer with many sprinkler heads for each valve and bed number.

Once again, I was loving my work and dealing with a long commute. The commute became even longer in May 2007 because I found a new partner, Mary Frances, and as was my pattern, I quickly decided she was the one. She felt the same way about me. When my lease expired, I moved in with her in the Woodlands which is north, but far to the east of Waller, farther than my apartment in Tomball. An hour became a fast commute; it could expand to an hour and a half if traffic was bad. At least I had weekends off.

The owner of Magnolia Gardens was Tommy Marek, the younger of Ralph's two suns. His older brother was head of Ralph's old company and the associated businesses. I'm sure Ralph knew he could never put Tommy in charge of anything, so he set him up with the nursery. Tommy lived on 320 magnificent acres in nearby Grimes County, where he maintained an exotic animal park and a phenomenal swimming pool at his more than elegant home. It was a great site for his annual company picnic and well over 100 were in attendance. Rumors abounded. He was on the verge of divorce, not so bad that, although both were devout Catholics, but that Tommy was sexually deviant and that was the source of the marital problems. Tommy rarely came by the nursery, and often didn't speak to me when he did.

One time however, I remember very well because he brought me some up-rooted seedlings of black gum trees, wanting me to plant and resuscitate them. I had a new member on my crew – they appeared out of nowhere from time to time – a young woman, very slender who cut her hair to look like a boy. I would describe her as gamine. She'd worked for me only a day or two and seemed like a good worker, fine to be around. As Tommy first approached me, he spied her and his eyes lit up. Within a few minutes, he had wandered off with her, and he returned about a half hour later to give me the saplings and chat for about one minute. He spent a half hour chatting up a 17-year-old girl, who looked like a slender boy, and five with me his propagation manager.

All Tommy knew about me was that I had been a favorite of his father. Ralph was still on Magnolia Gardens board of directors so I retained his distant support. The nursery's real inner circle consisted of the manager John, who was Tommy's cousin; John's two assistants; and a salesman named Ken. Manuel was in charge of all the shrubs and trees which took up most of the nursery space and provided about 60 percent of the nursery income and used 90 percent of the space. Propagation rooted most all of the shrubs and trees, and he grew on out in the nursery field. He was a sour individual with a smarmy way of ordering people around. Jimmy was more personable, but in a guarded way. He had been pesticide manager, but that had been passed on to Carey. Jimmy now was the Human Resource manager and general observer or spy for John. John was an imposing man with a very attractive daughter who was company secretary. Manuel and Jimmy were both diminutive and effeminate.

I mention Jimmy as a spy, and I don't really know that as truth, a good manager needs to know what's going on. Of course, a good manager needs to talk to all concerned whatever the issue, not just be a one-way conduit of information to the boss. Jimmy didn't do that. John spied and admitted he spied. The company office was in a large old house at the front of the drive. We managers and some support people had a separate office 100 yards away on the edge of our perennials acreage. Four or five of us sat facing a wall with computers and work stations. I had some cabinets full of testing equipment, rooting hormones and the such. Next to me sat Carey, who was in charge of all pesticide applications. He had some sort of debility that sapped his strength and restricted his observations. I didn't know how he kept his job. Manuel had a space, but never used it. I doubt he was computer literate. I rarely spent as much as two hours in the office, usually less. A corner was occupied by a young woman whose job to me was unclear, but I liked her. She didn't like John and seemed to be in trouble.

Shortly after I started, John installed cameras and speakers with which he could see and hear everything we were doing. If he was quiet, we had no idea that anyone was watching, but sometimes he would boom in and let us know he had the all-seeing eye. I often wondered later if I forgot the eye or ear and spoke poorly of Manuel or himself, both of whom I began to hold in contempt. The girl was fired not long after.

Actually, the contempt part didn't come until there was a big garden show in Dallas. Magnolia Gardens had a prominent display area in the hall, and my perennials were a big part of our display. The plants looked great, and it was announced that the perennials department had set a company record for sales. One of our experienced saleswomen told me that the perennials were the best she had ever seen. About 25 of us came up for the show, several of us with our partners to stay at the hotel. I had to work the show, but Mary Frances enjoyed a free day in Dallas. That evening Tommy put our contingent on a bus and took us to a great restaurant where we had a separate dining room. John and Tommy were pretty drunk. John kept pawing Manuel on the ride over and throughout the evening. Jimmy seemed to be cringing, and Tommy looked vacant as he focused on Jimmy. There were a few speeches about the team effort and company success. Manuel got special mention for his continued great work.

I don't recall being mentioned, but the food was really good, and I know I drank as well. The next day Mary Frances commented that John seemed to be an abusive gay man, that his helpers were submissives. Of course, that is what I believe as well. And no one would care if they were gay, but marrying that favoritism with business created an us versus them reality.

In December, Reagan came on board. Reagan had a master's degree in horticulture and had been teaching at Sam Houston State College. She was young and fully pregnant the day she started. She worked a couple of days, then had the baby returning to work a couple of weeks after that. She seemed to be from a local family, well off and probably friends with the Mareks. Given the prestige of her degree, I quickly surrendered the field, told John I would gladly work under her management. Although she knew little about actual horticulture, she helped the company a lot by scheduling crops and procuring seed or cuttings. I am not sure if she had the same initial task I had, but since she seemed to be suspicious of my management, I offered her four of my workers to do whatever she wanted in terms of plant management.

Slowly over about a month, she came to see that I was turning everything into gold, and she voluntarily returned the four to my management. Previously, I had placed an order for asparagus myrcofolides seedlings. Ming asparagus was an old favorite of mine that had never caught on at the nurseries. Soon I had 400 of them growing in three-gallon pots. By mid-spring they were selling well at a high price. My department was doing so well that management made plans for another huge perennials house.

Reagan then set out on an ambitious planting program that would more than fill us to the gills. She and I met with her salesman after we decided what crops we would grow. He apologized to us that he could only supply a small part of our order with plants. However, he could get us whatever we wanted in seed. In today's world of nurseries, most all of plant seeding is done in specialty greenhouses. The salesman assumed we would require pre-started plants. None of his customers seeded their own crops. Unbeknownst to him, the nursery had a seed-starting cabinet with humidity and heat control. There was also a complicated vacuum with variously drilled trays that would drop one seed into every cell of a flat and that in a short time one could pull from the cabinet dozens of bedding plant flats filled perfectly with sprouts of our cultivar.

I had never seen or used this equipment before, but planting by the moon, I got fabulous germination of everything that went in the chamber. Other seeds I sowed the old-fashioned way, thickly in a flat, and then taught my crew how to transplant seedlings. Our whole department became a proud and happy team as success followed success. Then Ken, the salesman, landed a coup. The major Texas grocery chain H.E.B. wanted to purchase 3,500 one-gallon pots per month featuring a different plant each month.

All the brass settled around a table to select plants for the first four months. For the first month, Nemesia was selected by H.E.B. We bought all the seed available and there was barely enough. The next two months were my suggestion. First were red geraniums and white geraniums under-planted with lobelia or alyssum, and the second was thunbergia, vining on a trellis. For the final month, plumbago was selected, a plant that was in our normal production schedule and that we did very well with.

I sowed nine flats of Nemesia to start everything off, and they sprouted well in the seed chamber, and then I brought them out on the bench so they could grow a week or so before we potted them up. So the very first night the seedlings were exposed, our company's feral cats came and shat and scratched in three of the flats. Those flats literally stank of cat urine and fewer than half the plants were salvageable. Our Nemesia, which needed to be thick, were now 25-30 percent thinner. What a bummer. I didn't like Nemesia anyway.

In the meantime, neither Reagan nor I had any idea exactly how long it would take to grow on the geranium cuttings. She went to our salesman who told her 13 weeks. She and I chatted, and wanting to be sure to have the flowers going good, timed them at 14 weeks. We had good take on the geranium cuttings and transplanted tiny lobelia and seeded the alyssum as we put them in gallons. The results were spectacular. We had several color combos; the white geranium and the blue crystal palace lobelia was the best and most common, but red-and-blue, red-and-white, even white-on-white, all in all 3,700 gorgeous gallons. Everyone was excited. They were beautiful.

Those of us caring for these plants, however, had great concern. It was spring, the sun and the heat growing daily. I knew that geraniums liked cool spring and that heat was a problem. I had been around geraniums for years, but never worked with numbers like these, and I knew they were a temperamental crop. My department did not include pesticide applications; those were all done by Carey, our afflicted pest manager. I don't know if he did a fungicide pre-soak at planting; I doubt it. Largely because of Reagan's ambitious cropping, we were full to the gills. I was barely able to move enough newly rooted plants outside so as to clear space for the geraniums in the propagation house. The prop house was allowed to get quite warm to root Pancho's heat-loving cuttings and it also had mist irrigation, the least desirable form for geraniums. We had to run the mist a long time to fully water the geraniums. I just did not have the people, or at first the will, to hand water them. Yet the geraniums were stunningly beautiful and the Nemesia were ordinary.

I went to Ken, the sales star who handled H.E.B., and begged him to send the geraniums first before the Nemesia and he turned me down flat.

"They've already sent their advertising and it says Nemesia," he justified.

I said when H.E.B.'s customers see these gorgeous geraniums, they were not going to care what they call them, that they would sell like hotcakes. Didn't matter. I was forced to keep the geraniums another month, and the lobelia and the alyssum had to be cut back after their flush. The geraniums began to yellow and succumb to anthracnose, and when shipping day finally came, they had to pull a lot of the groundcovers out of the gallons and on the whole the shipment was a disgrace. My loading dock friends told me that Ken was cursing and kicked one unfortunate anthracnose afflicted geranium "clear across the parking lot."

I was quickly hit with two more body blows. The worst resulted from an error Manuel had made before I even started there. He got the elevation of the 600-foot-by-300-foot new Stuppy range off by quite a bit. Our ventilating walls never fit perfectly and a big chunk at the southeast corner of the greenhouse was a basin and had no drainage. I remember noticing when I first got there that Patricia's crop in this sector was crap, but at first did not know why. Later when I took over the irrigation in that house and saw the flooding damage, I explained the problem to John, and he assigned Manuel to pull the landscape cloth and trench in drainage to repair this area. His repair did little to help. In this rebuilt space I put 2,200 critical Plumbagos. It was the only available place, or I would never have put them there. The salt got to them about a week before they would have been shipped.

Carey's last-minute spray had done nothing. I almost pulled that crop off, but when they became unsellable, the nursery had to order plants from elsewhere a straight-up cash loss of about $10,000. In the meantime, we were having to continually prune and disbud the Thunbergia, which should have gone out the door when they were stunning, which they were. Then at the end of April, they fired me.

When I drove up for work in the morning, John hailed me from the office door. Jimmy followed me into John's office. John was quite belligerent, saying, "I told you heads were going to roll." With Jimmy in tow, I gathered my things and said goodbye to my crew. Reagan was sad; she knew the truth. Pancho telephoned me two or three times over the next year. He told me that the ladies working with him had starting crying at the news. I know that he was sad as well. I so regret losing contact with him. Pancho occasionally called me with news for a couple of years, but I was too full of my new life to call him back. I am sorry my friend.

In the way of employment in Houston, I could find nothing. One company regularly advertised for people, but they were like Perfect Lawns of Austin, a grounds-keeping factory with money in mind; there was no horticulture in it. I did collect unemployment compensation. It turns out that "incompetence" as grounds for termination qualifies the ousted one for compensation. I am sure now that I would have eventually found something. I cannot remember if I had any savings, but Mary Frances was not going to pay my way.

So then at age 62, I took out my Social Security and added $808 a month to the unemployment benefits. Then my Texas sojourn ended rather suddenly after I saw an ad for a propagation manager at Park Hill Plants and Trees in Tahlequah, Oklahoma. It was a nursery I had visited as part of my school tree grant, ten years before. It was in lovely country in a historic city with a state university. All three of my daughters were living in the state. By the first of June, I was housed near my new job and earning $40,000 with a company truck.

Here's a winter view of the southeast corner of the lower house. The wheat sprouts are for me and the chickens, as back then I had a hand-cranked grass juicer. Most of the rest of the potted plants are perennials which I had kept outside in the fall for as long as possible while I cleared space inside.

Oxalis purpurea is a very prized perennial. It is slow and expensive to produce but sells like crazy. When I had full pots of one gallon like this, I could count on $7 a pot wholesale. I had a variety called Iron Cross, which was green with a distinct black cross on each leaflet. There is also a small-leaved yellow flowering oxalis that is a bothersome greenhouse weed, a red spider magnet.

I love asparagus. It's my new super-crop. It needs to be seed grown at a nursery. Thousands of plants can be grown outside in pots on pallets all year with little effort or expense. Forget those dried crowns; now you can plant a real asparagus bed.

27

Park Hill Nursery

I drove up the 500 miles from Houston to Tahlequah in late May 2008, and interviewed with the owner, Brian Berry; his sales manager, John; and another advisor of his. They kindly took me to lunch at the family country club and more or less assured me that they would take me on as propagation manager. That was confirmed in an email I got a few days later back in Houston. I loaded my Nissan hard body with my necessities and left other belongings behind with Mary Frances. This was not a breakup, and over the next year there were a lot of 1,000-mile round trips to Houston. I was very lucky to have the free use of a full-size F-150 that even with the six-cylinder motor, hauled and towed a lot of stuff. I stayed a couple of nights with Rachael and Rick in Sam's Point about 100 miles from Parkhill, but after that, Brian finished setting up a one-bedroom mobile home on a nice lot just a short distance from the nursery and a shorter walk down to the scenic Illinois River.

 I had 62 greenhouses and a staff of about 30 waiting for me my first day. How nice that the company carpenter had prepared a sign for my parking marked "Sand." My office, however, was a tiny room in the corner of an exceedingly ugly barn which also was the breakroom with tables for about fifty people. I was very disappointed to find no computer, no telephone, and no soil testing meters. It was 600 yards up a steep hill to the office where I had to go if I needed to use a computer, and I had no separate budget or means of buying except to get a purchase order. My new boss, Jay, the nursery manager who had been out haggling in divorce court when Bryan hired me, never even informed me of established accounts we had with some of our suppliers. He never liked me and ultimately got me fired. The drama leading up to that is another good story about being sucker-punched in a jaw dropping way.

Among the Mexicans, there was a kind of consternation that I was fluent in Spanish. This was unheard of in Oklahoma. At first, they liked it since the better communication made many things go much more smoothly. There were many fewer misunderstandings and do-overs. They could now speak with their manager about both work and personal matters. They also realized that my command of the language meant that I would know both them and their work far better than previous managers in Oklahoma. I was invited to and attended their barbacoas, their christenings, and their quinceaneras.

At first all went well. I enjoyed being a kind of patron. But there was a hidden hierarchy I was about to trip over. My assistant went by the name of Buzzy, and for a few months before my arrival, she was propagation's crew leader. She had no professional understanding of plants. Buzzy was missing her upper front teeth – I thought it might be from meth use – and she was neither bright nor hardworking. She missed several days of work when I first started, and I had to give her rides my very first week on the job. I saw that she lived in a trailer trash-type neighborhood, no trees or flowers planted there.

Buzzy did a lot of gossiping and was involved in all the planting ladies' emotional difficulties. I realize now that the only reason Buzzy had been elevated at all was because she spoke enough Spanish to give assignments and she got along with Jay. Jay probably liked the idea of running the department through Buzzy, and the two of them worked well together, possibly because she was no threat to him and would always concur whereas I might not. I could tell that Jay was irritated that I spoke Spanish. Also very important in the scheme of things were the two waterers, Basilio and Eugenia. Basilio and Eugenia were in-laws I believe. She was older, in her 30's, and the first to have learned the controller irrigation. She often reminded me that she had trained Basilio and should be in charge. But Basilio, at twenty-four, was far more aggressive and tended to dominate her. These two got to keep their own hours and earned lots of overtime, with much of the work in very pleasant evening hours. Both were from El Salvador. Later on when their rebellion began, I called them the "Enanos."

The whole crew seemed to be divided into cliques, and Basilio and Eugenia allied with Nicki, who was the crew leader for the important job of taking cuttings. Her crew was paid piecework, and she counted the thousands of cuttings and packed them in bundles of twenty-five. After she reported the total number of her crew's cuttings, I could, and did once, verify the count by the number of cuttings planted and put out on the benches under mist. So of course, there was always some discrepancy and while I said little, the fact that they knew I had counted was sobering to them; it went better for them before with Buzzy in charge. I, of course, decided after a few months that I needed to ditch the Buzz.

At the same time, a middle-aged Mexican woman who had the lowliest job in the department, began to chat intelligently with me about the propagation soil and her job of sticking the cuttings. Her name was Carmen Miranda or Maria Rodriguez, depending on which card she showed. She was actually legal and her married name was Rodriguez. She spoke little English.

Carmen had been working at the nearby Greenleaf Nursery on a planting crew, when a laden trailer had tipped over, crushing both her legs. After multiple surgeries she walked with difficulty, hence, her work seated and planting cuttings. I was able without opposition to raise her pay from $7.25 an hour to $8.50. I seemed to have broad control over raises and bonuses for piecework that always seemed counter to the lack of respect otherwise shown me by Jay and the office. It turned out to be a headache because every extra dollar I paid out was criticized later. I was surprised at the time that Jay and Brian didn't review my payroll decisions. I never felt comfortable making these decisions in a vacuum, but that's where they left me.

Carmen kept good records; more than half of my staff was eligible for various piecework bonuses which she verified, and she was excellent at mediating when there were employee complaints. The first big job I gave her was to manage our grafting effort which went on daily from December through February. Grafting was of huge importance to Park Hill nursery. The knives that were used cost $80 apiece.

I had a propagation list from the Sales Department. The break room was reorganized to accommodate the nursery's ten most-skilled grafters who could earn top dollar from this skilled piecework. Most came from outside my department; one among them became my friend, the garrulous Pedro. In addition to these ten, there were at least that many more in ancillary roles. Actually, all my employees were begging to be assigned to grafting. The break room was kept warm and toasty for the sake of the callousing stock, and it was nasty winter outside for the rest of us.

Two women waxed the newly fitted root and scion. Grafts were then carefully boxed in straw. They all had to be counted and quality work maintained. Helpers were needed to sort through the bundles of stock, cutting them to length for the grafters. There were trips to a walk-in cooler filled with these twigs, branches and roots of all the popular tree cultivars. By January, it was clear that we were doing a fantastic job. Everyone was happy except Basilio and Eugenia, who this year had nothing to do with it. Carmen hobbled around and managed it all with pride. Then a shocking series of events began to unfold which ultimately ended my years of earning a living at a job.

Buzzy left my Propagation Department and went to work in inventory. She could have stayed on with me at her same salary; I did not fire her. Unbeknownst to me this was a Cherokee vs. Anglo mentality and the Inventory Department became a hotbed of lies about Carmen and myself. Their manager, also Cherokee, had daily and convenient access to Bryan and Jay. I heard later from another manager that all these transfers were against longtime company policy. I only saw Bryan about once a month, and it took me that long to find out about the lies, stories and distortions.

Regardless, I damaged my own cause when Jay and the salesman John, who were responsible, failed to supply the wood needed to meet the numbers and cultivars they had given me. I felt let down when Jay did not form a crew to cut more scion wood which was available to us. Nor was I able on my own to organize such an effort. So the day before we were to run out of work and facing the difficulty perhaps of having to restart later, I took two fellows with me and we drove to the nearby nursery and into a large block of the trees that had been offered to us for cutting. Of course, they were Bradford pears. I should have been indicted for propagating these monstrosities. These trees we were cutting were probably unsellable, too large for anything but huge installations, and the people who did huge installations already knew Bradford pear was a dog. Per Jay, we had permission to cut scions.

I am sure I told the helpers more or less how to cut, but this was not something that should be done on the fly. I myself cut along with them, but was feeling rundown and ill, along with being propelled by my agitation. We cut a very large quantity of scion wood, which allowed us to go forward and complete Park Hill's most successful grafting effort ever. About a month later, the nursery owner sued Park Hill for the damage we caused, and it came to quite a few thousand dollars. Bryan came to me with Jay and John to tell me about it, and I replied, "I think it was Jay's fault."

This comment I am sure did little to help my relation with Jay, but Bryan replied, "I think so too."

If Jay had done his job this would not have happened, but of course, I was culpable and I knew this was on my head.

In the meantime, Basilio came and told me that they did not like Carmen. I had already made it clear that Basilio, Eugenia, and Nicki would remain in charge of the cuttings and the overtime involved. I presumed he understood that there was no overlap between his authority and Carmen's, which there wasn't. I was the one who now supervised his watering. I am certain he did not like that either.

Around this time, I began calling them the Enanos, after they committed small acts of rebellion, certain tiny derelictions of duty. Then Basilio announced that he was going to leave the company for a job in nearby Fort Gibson. I said fine. But first Jay, and then Bryan were afraid of losing him. I told them both that I was in no way concerned about losing his watering knowledge because I knew as much or more than Basilio, that it would be easy for me to train his replacement. They gave in to the young punk proving their distrust of me. Too bad I didn't have the guts to say, "He's fired." I believe my young watering expert pocketed an extra buck twenty-five an hour; he, of course, became immediately more disrespectful.

I did not have an EC meter. I finally got one in December. We had already bedded in our fall juniper cuttings, an important part of our overall production. I was still using the soil mix Jay gave me when I started. It called for six times too much lime because he used the suppliers' 4.2 pH number for the bark, but oh, sorry, it's actually 4.9 ph. Our giant mixing machine had broken down and was getting new guts, so we had to mix this already salty stuff on the ground with our front-end loader. When I finally checked our soil with the meter, I immediately dropped the calcium from 40 pounds to 10 pounds. Two weeks later, Jay came around and my mixing crew told me he said to drop the calcium down to 15 pounds, interesting, and cowardly, that he did not admit his mistake to me. I doubt if my soil crew told him I already dropped the calcium to ten, since they spoke little English. I wanted to get this tidbit to Brian, but did not succeed.

One windy day I got dressed up, in a suit coat and tie no less, to go and visit with the president of Northeastern State University about Park Hill's help with a landscaping project. On my way home, after my usual hours of work the weather became suddenly very cold and blustery. I decided to return to the nursery to check on things. I was alarmed to see us wide open with bitter blasts rolling over our young tender plants.

When they saw me, the Enanos jumped into the frenzy of door shutting for 62 greenhouses; I was closing and glaring as well. The next day I rebuked them, gave them an assignment of revising and cleaning sprinkler heads (needed), and told them I would do the watering. By days end they had fled. Eugenia was also taken in by inventory, and she became the source of even more lying and gossip. Basilio went with his devious-looking cousin to the main watering crew who did all the irrigation outside the Propagation Department. So Basilio now had his own company truck; he was working evenings and weekends with complete access to the entire nursery and all of prop which he knows like the back of his hand. I had my two new trainees who by now watered as well as Basilio ever had. My guys had plenty of contact with him and were appropriately friendly. They relayed what happened next as it occurred over the next few weeks.

Several, four or five, inlet lines to our sprinklers had been kicked so hard that they broke below ground. Hmm, we didn't do it. Each kick was a muddy two-hour repair while the plants in that greenhouse went without mist. I was very involved and got very wet and muddy. Then our breaker box was thrown and we had to scramble to figure that out. My guys didn't know where the box was and neither did I. Then our valves started plugging, and we spent about three days cleaning about sixty valves, each one took half an hour. We figured out that someone threw a bunch of chicken carcasses, bones, that were blasted through the ten-horsepower pump for our misters. It was some kind of glueish calcium sludge. I actually saw him from a distance driving off from a greenhouse whose controller was suddenly turned off. I kept Jay informed. He said I needed to set up some cameras to catch the culprit.

By the time Basilio left prop, two other things were clear: The entire juniper crop was toast and the high calcium per Jay's formula was the reason why. After the pear-cutting disaster, I'm not sure why they didn't fire me then. I can't remember anyone asking for my version of what happened. Somehow, they kept me on anyway, but ostracized. Rumors, Carmen supposedly, blabbed to everyone that she was going to get a new company truck like mine. Or that Carmen was giving bonuses only to her favorites. Ugly little things that went to Jay's ear.

Things that I didn't find out about till later. The leakers of the lies were Buzzy, her boss, Eugenia and Nicky.

The most blatant was after an urgent call for an inventory; all of the managers had to count one part or another. My duty was to give numbers and types of the roses I had. This I completed two full days before the deadline and entering the inventory office and finding nobody home, left it in plain view in the inbox on the desk.

At the meeting when Bryan inquired if inventory had gotten all the needed data, my tormentor replied, "All but propagation have turned in their inventories."

I heard about that a few days later as well and went to inventory to confront him, and as we spoke I saw that my rose count was still right there in plain sight in his inbox!

Carmen and I became very close. For me, as I am sure it has been and is for countless men of power, having a skilled female lieutenant who believes in you and supports you is very gratifying on more than one level. After the departure of Nicky and most of her crew of cutters – I don't even know where the four of them went – the shrunken remainder of our department became harmonious, happy and effective in their work. Pedro came in with his wife and quickly took over the cuttings. I had taken a big loss in experienced help and within a few short weeks, I had new people trained and old people who had left previously due to this same problem of cliques, now returned. Once again, my department was doing very well. In July, gasoline started disappearing from our three trucks. Jay wasn't interested in doing much about this either. So in August, I bought locking gas caps, then left on vacation to help move Mary Frances from Houston to the home we had bought two miles up the hill from the nursery. I was back home the Saturday morning before my report back to work on Monday when I got a frantic call from Carmen. Both of our work trucks, Carmen's and Pedro's, had been burned to the ground. When I got there, the smoldering carcasses were parked at the office in plain view from the street. At that moment I knew I was fired.

Jay waited till the following Friday after the checks had been handed out to inform me. Later all was cordial as I talked with Bryan, and there was no company resistance to my unemployment claim. I bought a bunch of trees and blueberries at wholesale and planted them at our new home. Bryan later became a friend of sorts, spending $200 once at my new Plants Alive, and I gave him a bunch of unsold cut flowers to decorate the tables of his restaurant, where I could never afford to eat. Within a few months of my ouster, Bryan's brother, Burl, took over the nursery and his first act was to fire Jay. I wrote to Burl about his prop department, but he was not interested in bringing me back and I don't blame him. Carmen also lost, first her position and then the pay raise I gave her, then later, was let go outright. I saw her from time to time in town. She was a resourceful woman and still working. She was cream constantly being remixed and then slowly rising to the top.

This is the plant I grew on trellises at Magnolia Gardens in Houston. Those were very pretty on the day they fired me, but not as pretty as this accidental planting vining up the greenhouse wall.

Blue tiger jaws: This is presumably a Faucaria, but I could not find it in my exotica. It would at odd intervals decorate itself with divine purple flowers. It was easy to propagate, and I grew hundreds of them. They stayed so small that I often used them as accent plants.

28

Plants Alive, 4712 Ridge St., Tahlequah, OK

I am not sure why Mary Frances chose to leave Houston, other than that her neighbor and best friend had just moved to Oregon. We had been together for just two years. They were mostly loving, but my sexual cravings were a lurking problem. She discovered the porn, of course, and she wasn't going along with it. We bought the Tahlequah house together in February 2009, but after I lost my job in August, she quite rightly took me off the title, particularly since she had made the down payment. My dollar contributions, while substantial from my point of view, were minimal in the big picture. I went along with her and became the no rights partner.

Nevertheless, I lived there with her in a comfortable three-bedroom, brick "Indian home" for ten years, longer than any other place in my life. These homes are scattered throughout Cherokee County, and once cognized, are easy to spot: They all have an attached one-car garage. The Cherokee Nation builds these homes for its citizens, who then pay the tribe back with extremely low payments. Our first good neighbor said she and her husband were paying $81 a month. Many of these homes are alone, set on land already owned by a tribal member.

Ours was in a 50-acre plot owned by the nation which had been divided into about 50 home sites. For those who stay in the homes paying $81 a month, it is like a forever rental since the principle is never paid. The tribe keeps restrictions on these residents including required mowing and upkeep and do not allow poultry or farm animals of any kind. They also re-roof and re-glass the houses as needed with no charge to the "owner." Homes in the sub-division were first built in 1970, and the last, ours among them, in 1995.

Some of the homes, four or five that I knew of, had been paid off by the original owners. Those properties now were no longer controlled or regulated by the tribe at all. Our house became the only one which allowed the weak and over-mown pasture to return to waving, wildflower-filled, Ozark meadow, and it did not take long. We also had chickens with rooster.

For the most part, we got along fine with our neighbors, more than one of whom burned plastic-ridden garbage with black plumes of smoke. I remember one horrible time when the wind took it right to our house. This was also the trashiest place I ever lived in. Plastic was everywhere, along with a few uncared for dogs that usually perished early. I saw a big pure bred-looking black lab starve to death near the house. Thank goodness our rooster's crowing didn't seem to bother anyone.

Ours was a very nice property, 1.16 acres with some deep soil on the low side of a moderate slope. The west edge of the lot was a wonderful fencerow. Fifty-foot cherry trees led the way up, but the space below them was crowded with sassafras, privet, hackberry, half-wild pears and sumac, all of it woven together with greenbrier or mustang grape. The fencerow very much wanted to spread beyond its narrow belt and where it wasn't mowed it did advance. Even when confined, the fencerow used lots of water and their roots easily extended 25 feet into the backyards of everyone living on the west side of Ridge Street, mowed or not. Not foreseeing the effect of that and believing I could grow plants anywhere, I said to myself, "A bed would be pretty here," and I did work hard on that bed which was set on top of and among those same roots. The hollies and barberry I planted there, with pickaxe, have very slowly gotten established over the years and are now quite pretty, but the bed was always dry. My flowers were OK, but they never really thrived. Meanwhile, the fencerow gang, looking well fed and watered, smiled down on my frustration and wilted coneflowers.

The lot had only three trees of note when I moved in on my birthday February 2, 2009. On the east side front entry of the house, a neighbor had planted two surplus silver maples from his nursery. Nice to have the shade and ambience of these now 40-foot tall trees, but I can think of at least a dozen better trees that don't rain dead branches on the ground or forever sucker. Out in the field 150 feet from the house was a single old redbud. It was large and still shapely, but I pruned a lot of deadwood out of it and thought it might die soon. Instead over the next ten years, it just kept getting more and more beautiful. Then the summer after I left, a squall tore the old tree in half, revealing the rot inside, common with old redbuds. This one might yet sucker around the rim of its old trunk for a hundred more years.

Before we bought it, the entire lot had been mown per tribal policy. It wasn't what I called turf, however. About half of it was Bermuda grass, but there was also Johnson grass and a full complement of fescues, clovers, lespedeza, and forbs. It all looked unhappy that first spring, but then I quit mowing, except for pathways and small plots in front and back. All the mowing I did do was with the deck set to its maximum height. I steered the mower around natural patches of yarrow and Dutch white clover. After that the wildflowers just seemed to jump out of the soil.

I planted a lot of fruit trees that first spring. This land had been an orchard in a previous life. I sure wish I knew then what I know now! The apples, plums and cherries did poorly, although one apple cultivar was starting to bear a little without being sprayed when I left. Both plum trees died right away – graft failure, I suppose. The two peaches and the nectarine took off with good form and great promise. The third year I had great fruit set on one peach. I did all I could with dormant oil and a few organic sulfur sprays, but the 100 pounds of peaches I picked dwindled to 35 pounds by the time I cut the rot out and froze the good.

These three trees quickly grew into giants which required more massive pruning than I was able to do. I got more good smoking wood than I could ever use, but it was not a good trade-off. Over ten years I never got a single unblemished tasty fruit from any of them. One of the peaches blew apart in a windstorm just before I left in the spring of 2019, leaving a 40-foot circle of devastation. It required hours and hours of chainsawing and dragging. Thank goodness my pears and blueberries did well. I planted five pears. Two were dogs, a red Seckel and some other sweet pear, but the three others were Asian types, one being grafted from the T-town apple pears. To get the scions, I had to return to Tulsa and cut from the old trees I had last seen thirteen years earlier. Carmen's crew grafted about fifty of these trees, although after I left nobody at the nursery would know what they were.

T-Town pears? These trees yielded so well that I was busy from late summer drying them, making pear butter, and pear pies. There were over 100 pounds of fruit seven straight years, except for one crop lost to spring freeze.

T-town apple pear tree from scion wood gathered in Greenwood and grafted at Park Hill nursery ten years old.

Three blueberry bushes yielded well every year. The one year I kept track there were 19 pints over a long season of bearing, about $100 worth of fresh, organic blueberries. By my third year there, I was harvesting a lot of asparagus, all of which I had grown from seed then transplanted. In the garden area where I mowed pathways and did shifting agriculture, I allowed certain species to thrive. Those thrivers included plaintain, poke, lamb's quarters, castor beans, bedstraw, yellow and curly dock, and a Jamaican amaranth that grew eight feet tall.

The amaranth came from seed smuggled in by a friend. In Jamaica it's the green cooked as callalou. It remained good to eat well beyond tender. Most of them grew to eight feet tall and five feet wide, an excellent green food.

I continued planting trees, shrubs and perennials, and nature did the same. Pine trees, magnolia, slippery elm, crepe myrtle, lilac, Japanese maples, barberries, azaleas, horseradish, goji berries, kiwis, filberts, mock orange, more blueberries, blackberries, and grapes all planted and established by hand. My two grape vines clambered over the chain link fence, a former clothesline, and a rickety arbor extension to cover an area 30 by 40 feet. Underneath that shade, I rooted my cuttings in summer, and the annual late summer harvest of 20 pounds of organic grapes were a welcome reward for my part in providing the vines with space, water, and support.

Nature in the meantime kept planting as well. Many were welcome: fragrant honeysuckle, healthful mint, plantain, bedstraw, beautiful violets, more sassafras, mimosa, sumac, and lespedeza. Many ornamentals were planted from my neighbor's nursery windfalls. There was a collection of Roses of Sharon, a wonderful old-style white bridal veil spirea, a fragrant lilac, a yellow barked maple, a Shantung maple, a sunrise redbud, a trellis with kiwi, a cluster of goji berry I grew from health food berries, two hazelnuts I purchased small and grew to ten feet tall, two large clumps of elderberry that were one too many.

On many sunny winter days, the sky would be darkened with this shit. It kept the mornings cold and just when the plants in the greenhouse most wanted warm sun.

The yield from the vegetable gardens I had crammed in among all this was ordinary, but welcome. I did well with peppers, asparagus, cabbages, beans and potatoes, less so with my tomatoes and cucurbits, which were damaged more than one year by the toxic manure I had unwittingly hauled for garden and greenhouse. Greens came and went, but along with the peppers there were lots of fall tomatoes; of course, they were all small. I loved watching the hazelnuts with their weird winter catkins. I harvested a grocery bag of their filberts, but they were so small it made no sense to

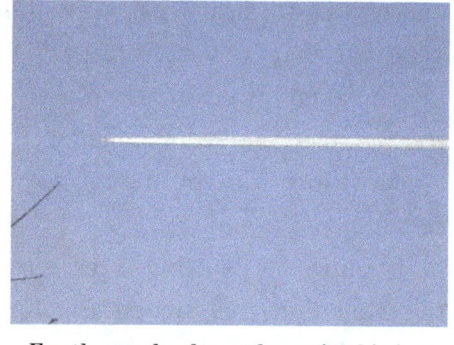

For those who do not know it, this is a chemtrail. It has nothing to do with exhaust from jet engines or atmospheric conditions. If you watch it you can see it billowing out exactly like crop dusting but high up. I believe this is an operation conducted by the Rockefeller-involved deep state. It's been captured and analyzed by brave citizens. There are lots of aluminum and barium, both bad. Some have suggested that it is coal flyash caught by the power plant scrubbers.

husk them. The kiwi vines were very large and on the verge of a good crop, weather permitting. I also expected my first goji berry crop from the two well-established plants. My friend Lane from Texas drove through and said it was like magic being there. My friend, Jesse Vincent, an indigenous herbalist, often came to visit. Starting from my little acre, he and I replanted the world. We admired every plant together, every drop of water, and every beam of sunlight. Then after a difficult decision in November, I left at the end of May 2019. The ten years in Tahlequah were the only ones in my life that allowed me to plant a tiny piece of earth and work with nature and see it grow into something that I could cherish.

"Statice" and red monarda. This photo includes a bumblebee. We had no, zero, honeybees at our house. We were too close to my old nursery and their bee-killing pesticides.

29

Plants Alive III

Over the final ten years of my career, I grew more different plants and cultivars than I had ever grown before. I had worked in huge nurseries, but had actually grown fewer varieties. This expansion of experience combined with the many years I had put in before, caused a blooming of knowledge that people recognize in me and some of which I hope to put out there in words. Over the ten years, plants I grew from seed or cutting sold for about $20,000. By the last two years, I had secured myself $5,000 annually in sales and my expenses were low. Half of all the soil I used came from leaf mold, composted manures or composted sewage sludge. I had to pay for heat, which was about $200, clay pots $800, peat/perlite, minerals $200, fuel $200, say $3,500 net. Good thing I have my government pension. If I wanted to make a real living as a wholesaler, I would need 20 of these greenhouses or about 10,000 square feet, which I believe to be the most a single grower can care for.

My first five years of business, I sold almost all my plants at the Tahlequah Farmers Market. Bless their hearts, I enjoyed being there and met hundreds of good people, including my "brother-friend" Jesse. He changed my life and is still cheering me on after eight years. I sold $1,000 my first year, built up to $3,000, then dropped to $1,500 after my plants were heavily damaged by toxic manure.

But it was so much work. There were only four hours a week for sales. Friday evening and Saturday early, I had to load my big truck with all the plants that were ready for sale, but worse, was loading the tables I had devised to set them on. Six cinder blocks, two ten-foot two-by-fours, and three heavy pallets loaded twice, unloaded twice. There was also a 10-by-10-foot canopy to set up. At 8 a.m., I had to unload and set-up at the market only to tear it all down reload, bring the unsold plants back home at noon, beat up from the wind and handling and just as exhausted as I was.

As you can see in this photo, I dug out the aisles to sink the greenhouse and create raised beds which either grew veggies or were covered with pallets for container growing.

Sales were disappointing, my best days were $200, and I averaged about $2,000 a year with them. I was surprised and disappointed that my excellent cut flowers were a sales dud. Then in 2016, I found the Ozark Natural Foods Co-op in Fayetteville, Arkansas, sixty-two miles east on US 62. I began wholesaling to their farm and garden department, and they took all the succulents I could grow and also bought my houseplants, hanging baskets, trees and perennial flowers. It was a relief to back away from bedding plants, which everyone else grew in the spring. That gave me space to pot up my fall-sown perennials, which in turn soon went outside to finish and bloom in summer. I also began propagating from our very nice collection of succulents so that before summer's heat, I began bi-weekly trips to the Co-op, carrying $300-400 each ride. Everything I took was sold, and my yearly sales more than doubled.

During these years, I was able to grow and learn the culture of an astonishing number of species I grew 99 percent of what I sold from seed, bulbs, cuttings or divisions. These included medicinal annuals and perennials, flowering annuals and perennials, food and herbal crops, shrubs and trees. I never learned grafting well enough to call myself a complete nurseryman. That's a rare club!

This is the final form of Plants Alive just before its dismantling in 2019.

In the spring of 2019, I was a much-favored vendor at the Muskogee Herb Fest and sold $893, my single one-day record. I also, thanks to Jesse's introductions, spoke to the 30 members of the Muskogee Master Gardeners chapter, and they installed me as a speaker at the state-wide Master Gardener's annual conference. I got a $50 check from Oklahoma State University, and Paul James, the keynoter, purportedly got $1,000 for the same conference and less time. Twenty-five years earlier, Paul had done one of his popular gardening shows on my organic work at UCT. He remembered me and seemed curious when he came by my plant booth. Never mind, the attendees of my speeches got protein, while Paul only offered pablum. My star was ascending as they say; it's fallen back to Earth since then.

I can only say that I loved being Plants Alive again and growing these plants. The nursery was small, 530 square feet of greenhouse with about the same space growing plants on pallets outside in summer. Watering rarely took me an hour, and while I might spend a half day with the nursery or even all day in planting season, the work was pleasantly physical, but rarely strained. Best of all, I could almost always do critical things early and take the rest of the day off. Nothing was ever rushed, except getting back from Tulsa if it turned cold and the greenhouse was unbuttoned.

I heated with a combination of wood stove and propane heater, and after seven years with wood, the heater burned out and will not be replaced. The little buddy heaters did one whole winter for about $120. Critical to the process is not to overheat in winter. I often dropped to 33 degrees Fahrenheit in the early morning. If the nighttime low was say 30 degrees outside, I didn't heat. With no heat most nights, I could hold about seven degrees above ambient. In the winter and cold spring, I ventilated at 70 degrees, and I could ventilate from all around

Thunbergia, Black-eyed Susan, Crown of Thorns, Euphorbia splendens.

the house. People assume that heating the greenhouse is the most important aspect of its design and function. Then they heat at night to 55 degrees or whatever, and in the daytime, temps go through the roof, 100 degrees even. Inexperienced growers think this is normal and good, never mind the red spider and white fly, who are the primary beneficiaries of this ignorant thinking.

As I mentioned, I never had to spend more than an hour watering all my container plants, but more was required to fertilize. I set drums on platforms and filled them with compost, nettle, or comfrey tea; then when it smelled right, I fertilized organically with gravity flow. I hauled manure, leaves, and mulch in a huge decrepit Chevrolet pickup, composted the manure and set the leaves in corrals waiting 30 months for leaf mold to form. These I screened and used in my potting soil, along with peat moss, perlite, and organic minerals. Very few if any nurseries make all their own soil. I came close and could do so even now if I had access to truckloads of clean cow, goat, sheep or horse manure. Then if I hauled about 100 bags of leaves from street sides, I could more than replace the Canadian peat.

I found free used pots and never bought a plastic pot or sheet. I did, however, accept a big chunk of 6-millimeter poly from the University of Arkansas, which I used as needed over five years till I finally had to send it to the landfill. If I had been operating a bigger nursery and needing to make a profit, I would have found it necessary to buy containers which are packaged and priced to serve very large nurseries. Of course, plastic is disgusting in all its forms.

Rose of Sharons, monarda, day lilies. Everybody loved the pale yellow daylilies. The red Monarda or bee balm is another great perennial. I propagated lots of it.

I planted my largest crop, the succulents, in clay pots I bought at the box stores. These four-inch plants sold for $4.50 to $6 in a pot that cost me a dollar. In order to sell $2,000 worth of Thunbergia, black-eyed Susan, Crown of Thorns, Euphorbia splendens, succulent plants, I had to maintain stock plants of roughly equal value. These were my oldest plants, some of them worth a great deal of money as really large specimens of succulent plants are rarely available on the market. Many of these more beautiful plants became so important to me that they remain in my memory as long-lost friends.

SEPTEMBER P6 VENDOR OF THE MONTH

PLANTS ALIVE
Crista McKinzie, *Homestead Manager*

Note: This little article about my last nursery was written by the farm manager of the Ozark Natural Foods Co-operative. P6 is a term they use for a small vendor whom the co-op has inspected and qualifies for them as organic without the government certificate.

Plants Alive is our P6 vendor for the month of September. Plants Alive, run by Sand Mueller, provides the co-op with garden ornamentals, annuals, perennials, hanging baskets, and succulents from his wonderful collection. Every couple of weeks, Sand loads as many plants as his vehicle will hold and makes the trek to the co-op to sell them. As I watch him unload and he starts to tell me about the plants he has brought, I can tell these plants were grown with much love and attention. It seems that some of them are even hard for him to part with. I always joke that waiting for Sand to deliver is like waiting for Santa to come. I can hardly wait to see what goodies he is bringing.

Sand, whose long career in horticulture began in 1971, has built a small but unique greenhouse at his home in Tahlequah, Oklahoma. The thrust of his career has always been towards organic and sustainable growing, and he has read all the great pioneers of organic agriculture, including Sir Albert Howard, Friend Sykes, F.H. King, *The Secret Life of Plants*, and the *One Straw Revolution*. For the past five years, Plants Alive has been a popular fixture at the Tahlequah Farmer's Market.

Thank you, Sand, for all the beautiful plants!

Crista McKinzie, Homestead Manager

To build an azalea bed: In a good site, dig a substantial trench, heap the dirt around the sides, fill with bales of Canadian sphagnum peat. Turn water on low, and in your skivvies work the peat with your fingers and running water to saturate the peat. Set in azaleas 3' apart, three for every two bales, or if big plants, one per bale. Water to establish, remember water just the original rootball until new roots enter the soil matrix.

30

The Last Days of a Horticulturalist

And His Desire to Talk about It

The 1970s were boom years in horticulture and with the new greenhouses came lots of pests. I built my first Plants Alive greenhouse in 1973 in Espanola, NM, and operated organically, using no pesticides, but my greenhouse was small potatoes. Back then the chemical industry was churning out new pesticides, each more deadly than the last.

My contact with them began in 1977, when I finally gave up on my business. Over the next ten years, I used just about all of them. When spraying, there was always contact with the poisons, even when fully garbed. If you could smell it, you were getting a dose. My very first job at Payne's Nursery was to "Just put a teaspoonful in each pot." The smelly granular product I was applying without a mask – the customers might notice – turned out to be a formulation of one of the most toxic poisons ever allowed. This product, methyl isocyanate was manufactured by Union Carbide in Bhopal, India, where between 3,700 and 16,000 people died when the factory had a leak. Over half a million were injured.

In 1978 working at Breiters in Elmhurst, I was tasked to spray the whole greenhouse which included a lot of hanging baskets. I was covered in quality gear, but I erred by putting my sleeves inside my gloves. The poison, Meta-systox-R, had a dangerous Lethal Dose 50 of 65 and smelled hideously. The spray rolled into my gloves, off my slicker, and the poison soaked my hands. It's hard to notice much as the body is hot and sweaty wet all over. I spiked a 104-degree fever that night, did not go to doctor or ER. Marsha did a more or less continuous rubdown with isopropyl alcohol. I seem to recall feeling fine and going to work the next morning.

My pesticide exposure peaked between 1985 and 1987, when I was working in Mexico. I sprayed really toxic substances on carnations and other flowers for the florist trade. Once standing near the mixing tank and dressed for spraying, I felt a sudden wave of being enveloped in pesticide dust – Lannate had an LD 50 of 17, if you know what I mean. I instantly took all my clothes off and had my assistant hose me down. It passed.

After we left Mexico, I landed in 1990 a plum state job as the grounds supervisor at a new showplace campus in Tulsa. Within a year after securing the Tulsa Zoo manure, I declared it to be an organic campus. I made one exception; we used a chemical declared to be completely safe and biodegradable. It was new to me; Roundup was its trade name. It was very effective; nobody I knew took precautions of it.

At about that same time, I began having frequent and serious urges to pee. Working all day outside I became furtive, peeing behind trees, truck, hedges, and buildings. My bladder was reacting violently to the rain of toxic chemicals being released into my system in the years after the exposures. By 2010, I had detectable blood in my urine. As became a habit later on, I was slow to get medical assistance.

In February 2012, my urologist cut out and sucked out a tumor the size of a tennis ball, with lots of "kelp," from my bladder. The lab found it to be highly malignant. At stage 1-D, it was just one step away from bursting through the bladder. If that happened the cancer would have access to the rest of me; goody.

I came across two relevant facts at this same time. One was that pesticide exposure like mine often led to bladder cancer. The other was that bladder cancer kept coming back, more than any other. My urologist had a treatment protocol which called for four consecutive surgeries three months apart; the next year, two surgeries six months apart and so on.

So accordingly, in July 2012, I had my second surgery, delayed a couple of months to allow my Part B to kick in. After the surgery, the doctor stated that he was impressed and amazed, that there was almost no cancer.

How did this come about? Bearing in mind that the AMA says there is essentially nothing a cancer victim can do to treat his own illness, save put the professionals in charge, let me now chronicle the rebellion I took in response to my cancer.

I acquired a reverse osmosis water filter to take out fluoride, and then began following the Johanna Budwig Protocol. Budwig was a brilliant PhD biochemist, who in the 1930's was a young protege of Otto Warburg, the Nobel-winning father of cancer research. Warburg is out of favor with current treatment modalities as he emphasized environment, not genetics, as a cause of cancer. In the 1940's, Budwig became convinced that the nutrient that would heal cancer was an Omega-balanced fatty acid. And her chosen fat was flax oil. Fats – she became the acknowledged world expert on fats – cannot pass through the cell membrane; they can only be stored. So taking any oil, like fish oil, does little good. She needed a carrier, a high-sulfur protein was what she sought.

That was quark. Quark is cottage cheese. Actually, quark is Nature's prototypical cheese. It was discovered milliards ago when the earliest herders stored their milk in amphorae.

It is exactly the curds part of our favorite nurse nursery rhymes and forms from raw milk allowed to separate. Fermentation is started by enzymes found only in raw milk, so pasteurization ended the sale of quark, and that's why you never heard of it. I pour my curds and the whey into a pillowcase set in a big pot or bucket, then tie it up and hang to drip into the bucket. One photo shows the quark from inside the pillow case and the whey as well.

The quark is emulsified. An immersion blender emulsifes the flax oil; its yellow color disappears into the blend. Ah hah, now I remember the emulsions we made in high school chemistry. The Omega rich fats are now fully soluble and pass readily through the cell membrane. To the emulsion is  added fresh ground flax seed, blended fruit, nuts, seeds, cinnamon, whatever you like. It is very a wholesome achievement.

So in addition to the dietary emulsion, Budwig insisted upon an organic vegetarian diet and on sunbathing for vitamin D. There was a proprietary oil, quite expensive, which she also called for. I read that Deepak Chopra recommended sesame oil, and so I used that, and usually put it on for sunbathing. Over a 40-year period starting in 1960, Budwig cured 90 percent of the 4,000 patients she saw, most of whom were in extremis. She has a Wikipedia entry and her protocol is described at www.healingcancernaturally.com

 This cure rate is no longer achieved in her absence, but I have followed it with total confidence from the onset of my illness. I cannot say that I adhered to the protocol as I should have, but have stayed close to the heart of it.

Friends from Maine and Canada sent me some big chunks of chaga mushroom. Chaga is a very hard, shelf mushroom found on

birch trees that circle the globe in northern latitudes. I researched the preparation of it and began by soaking the piece overnight in the best water. Then I carefully chopped/cut it into granola-sized pieces. These required both a long boil extract and an alcohol tincture to release the full healing benefit. All water with even a little exposure to chaga would turn brown and was eagerly consumed. It just reeked of healthiness. Chaga was the treatment given Russian cancer victims at the time Alexandre Solzhenitsyn described it in his novel *Cancer Ward*, and has legendary qualities ascribed to it by northern indigenous people.

When I consumed the tincture, I sometimes got quite tipsy from the jiggerful of tincture. I also took a $150 Hungarian fermented wheat formula for one month and began Tai chi classes in our town. I went to a chiropractor who assured me that without his alignments, my energy channels would be impeded along with my healing.

After my second surgery, such an easy one with no cancer, I did not follow my urologist's schedule, but after a year, family pressure wanted me to go back, so I did. My urologist was clearly irritated that I was not following his timetable. But there was no blood in my urine, because if I did still have cancer, it would now be low grade. He then insisted on looking inside with his periscope. He detected some cancer and said I required another surgery. I should have left then. This time he wanted to put a chemoagent in.

Although I didn't know it, this was another clue. There was only superficial cancer. Chemo is not indicated over fresh wounds because the healing is retarded. He wanted the chemo for tissue that did not yet have a visible outbreak. Standard medical practice will never tell you that bladder cancer can be a slow-growing chronic disease that can be well tolerated till the end from some other cause.

Recovery from the third surgery was noticeably more extended and painful than from the previous one, and since I detected a certain hostility in my urologist, I cut myself off entirely from AMA medicine and began a study and vertical practice of alternative medicine. By vertical, I mean that I began growing, preparing, and consuming many different medicinal plants. It was an effort that few could duplicate. In my favor was that I was retired and could do what I wanted. I was a horticulturist skilled in plant propagation with a nursery and garden, so I had the means and methods to attempt it.

I took for my first effort the production of Rene Caisse's Essiac formula. It included four herbs: burdock, slippery elm, turkey rhubarb, and sheep sorrel, and I ordered their seeds. Although sheep sorrel grows in fields just a few miles from my house, I never arrived at the right time to harvest any. And while my seeds did sprout, I never succeeded in establishing a stand. I also sprouted the turkey rhubarb, but that plant faded in hot weather. Burdock grew well for me, and I still have some dried root years later. I got perfect germination from slippery elm seeds and grew them in 5-gallon buckets, sold some and planted others. My best are getting towards 20 feet tall after six years. So my homegrown Essiac cure, not successful.

It did, however, open for me a five-year binge of growing flora of medicinal value. There were about 40 different species of medicinal plants I grew from seed or cuttings, and most of them I planted in our garden. I came to know them well. In addition, I harvested and used plantain, wild cherry, dock, mullein, sassafras, bedstraw, yarrow and poke already growing on my acre. I based all my selections on plants referenced in my three books on plant medicines or described on the internet. In fact, most of the direct knowledge I have of the effects of these medicines is obscure to me, hidden in the inner intelligence of my body. My pineal gland, gateway to this inner knowledge, is calcified with fluoride stones, or so they say.

For the next three years, this was my only treatment. Then in 2016, I was nagged by pain in or near my left kidney. I went to a different urologist in Muskogee and took every test that Part B would pay for. All my urinary functions were good. The urologist, and I liked him, let me see the inside of my bladder on his monitor, and there were only three tiny patches of very superficial cancer. I was egotistically impressed and believed I had stifled this virulent cancer. Nor did my doctor argue that I should have surgery.

Then suddenly, he told me that my PSA was sky high, 61, and that he was concerned I had aggressive prostate cancer. He wanted to do a biopsy. I told him I needed to think about it. Then, on the internet, I found every reason not to go along with him: "PSA test unreliable" and "PSA scores higher after digital exams." I felt fine, everything still worked, how could I have anything other than benign hyperplasia with my vigorous body supported by all I was doing?

So against his medical advice I returned to my own medicine. Being poor, I even dropped my Part B. Over the next three years, I saved about $5,000 in this way. I also began falling away from Johanna Budwig; raw milk was becoming harder to get, all-organic, vegetarian became a lot of organic vegetables, occasional meat, sugar – I even began smoking cigarettes with friends and then on my own. There was disharmony in my relationship at home. So inside my body, mostly unfelt prostate cancer grew and marched into my bones.

In January 2019, my urinary functions gave out. I became incontinent, wetting my bed at night, then bloating painfully with retained urine. Weak and untreated, approaching death, I returned to the urologist and was drained with a Foley catheter. Since then, now 15 months later, I have been continuously catheterized, and that has led to much pain, many trips to emergency rooms, and constant risk of infection. The pain and ER visits were caused by unpredictable clogging of the catheters by both sediment and blood clots. When they are clogged, the pain of being unable to pee comes in increasing waves of agony. I have learned to head to the ER at first certainty of the clog which I can ascertain when my syringe will not push water back through the catheter into the bladder.

As a result of my urinary situation, I returned to the urologists. Since I had moved to Rachael and Rick's property in the countryside near Oklahoma City, I found a new one there: the very old and kindly Homer Claude Hyde. Dr. Hyde made it clear to me that I had metastasized prostate cancer. In fact, he said I was exceptional, that he could hardly believe I was standing in front of him. He had noted my prostate as cancerous from his digital exam, but what shocked him was how high my PSA was, over 4,000! He said I needed to see an oncologist, but did mention that there was an operation that could help me, a subcapsular orchiectomy. It was a castration that left the testicular sacs but cut out the cells that create testosterone. The reason for this is that once prostate cancer heads for the bones and other parts of the body, it is propelled by testosterone.

My operation was delayed by a sudden need to move from near Oklahoma City to Tulsa and with that the need for a new urologist to accept me and get all my records. Finally, on October 17, the surgery was done. A bone scan taken just before the surgery labeled me Stage 4 prostate cancer. I think that is the stage where they say they no longer hope to cure you. However, with care, people in my condition have lasted as long as eleven more years. My new, and first oncologist, said he was sure I would last at least three more years. Apparently, the loss of testosterone can really set the cancer back, but there can be nasty cells that are able to keep going and no longer need my manhood. Where I land on that scale will likely determine my end time.

I face one more surgery now which will open my urethra and scalp off some new cancer on the inside of my bladder. After three more years, my pride in trivializing the bladder cancer has become another operation. Fortunately, the two will be done together. After that, as the oncologist says, "We don't have to do anything, just watch."

December 9, I had that surgery which was a bladder resect and a trans-urethral resection of the prostate or Turp. Six or seven hours later, they had to wheel me back in for a second surgery to cauterize excess bleeding. I spent the next three nights and days in the hospital.

The following is a concise description of my state on Dec. 28. The bladder cancer was low grade, very much an affirmation of my resilience and surprising to Dr. McGeady. Also surprising, my prostate was not as strangling of my urethra as he had expected. The prostate was, of course, highly malignant. I still do not know what my PSA is; it apparently can take months to settle down. So this heavy-duty operation, from which I am still recovering three weeks later, did very little to help my condition. It seems likely after the first test that I will not be able to urinate on my own, but will have to continue with a catheter. Dr. McGeady told me that my entire bladder seemed damaged even though the cancer was not that serious. I wonder if the chemo back in 2013 had ruined it.

The PSA blood test came in; the number remains high even though the testosterone is low, 893.

Both my doctors suddenly suggested I do chemo to add a year to my life. When the PSA doesn't drop, it indicates that the metastasized cancer no longer needs the hormone, that the jolt to the cancer and years of remission we had hoped for are not going to happen. Maybe I have only months to live, and yes, I am experiencing a little more pain in my pelvis and in my upper arm. And yes, I have begun taking the tiny opioid pills, two or three per day, as I pass the one-year anniversary of my bladder failure and catheterization. And no, I will not be taking any chemotherapy.

July 2020 update: After a surprising call from oncologist, I went back in. He wanted more blood work, which came back at 154, a very belated good sign. But then the next number came in at 281, so it is unlikely that I have any salvation there. And to end my tale as the nasty summer of 2020 approaches fall, my doctors have a kind of renewed hope for me. I am feeling no worse and am no more incapacitated than I was a year ago. My urologist, I call him Jim, now thinks some expensive hormone pills can continue to drop my PSA. All three of my doctors expect me to last a while yet anyway, and so my opportunity to spend time considering my own death is extended.

Here I add the very day this goes to my editor, that my urologist called me out of the blue with a "new hormone" of great dollar value is being offered to me free of charge and the treatment may well drop my PSA to below 10 or 4, or near zero. The nurse and doctor involved are sincerely caring, and I never considered refusing their help, just bi-weekly blood draws to see my numbers and guard my liver. Got to take prednisone too, since my adrenals are gone. At the same time, I have met remarkable healers whose advice and foods have kept my decline very slow from before any pharmacological treatment. Part of this is a resonant healing device that relates to the Royal Rife machine, but cheaper and more modern. I am very conflicted and to conclude my story by saying tomorrow I have an appointment that will give me all the numbers and where I can ask all my several questions.

I want to say this about being critically afflicted with cancer. We all must die. I am now 74, and if I die soon it will not be a long life, but it will have been a full one. Since my affliction began, I have suffered much pain and about half my physical functioning is lost. I have a hard time putting on socks or tying shoes. I find it difficult to pick up anything from the ground. I walk a little or a lot slower, depending upon whether my catheter is bothering me. However, I can still cook, clean, drive, shop and take care of my pets and my plants. My illness has eliminated the possibility of work and daily activity as I always experienced it before. This has freed me to write, read and communicate. For years I had been saying I want to "do less and talk about it more." As you see, I have a lot to say, as well as my children. Below is something my daughter Arena wrote:

My Father

As long as I can remember, dad has turned to the I Ching for spiritual guidance. During times of change or uncertainty, he pulls out the tattered volume with torn pages and produces three coins. The coins are held and meditated over with a specific question in mind. Finally, the coins are tossed six times forming a divination pattern.

One late summer morning, beating the heat of day, my siblings and I drove 90 minutes into the heart of Oklahoma to talk to our ill and aging dad about moving to Tulsa. He was trying to homestead on our sister's property, but there were growing conflicts and problems. When we arrived, we found him in the yard, dressed all in black, moving through Qigong motions, his dog, Lucky, ever nearby. Long arms pushed and pulled, his body flowed, and the slow huffing and puffing made him appear short of breath.

Dad had a hospital bracelet on his arm from a few weeks prior, so I asked him about it. He said that it served as an amulet, as amulets have served ancient peoples for thousands of years. Also, his mother's practicality surfacing, he said it helped him remember the last time he had his catheter changed. Finally, he winked, his blue eyes managed a sparkle.

"It buys me credibility when I use the motorized carts at the grocery store."

He laughed about being judgmental in the past when he saw others using the carts. He ushered us inside his small shanty and told us what the I Ching had revealed to him. The ancient divination had shown my father and my brother-in-law standing on opposite hills or mounds, holding clubs for three years.

Today, while searching for and making calls to low-income senior independent living facilities for dad, I heard a song. It was a song that makes me cry on a good day. Although, it was a cover, I felt the emotion.

> *See I remember we were driving, driving in your car*
> *The speed so fast I felt like I was drunk*
> *City lights lay out before us*

Memories of growing up with my dad came into focus. Nighttime in our old blue Ford Econoline van, sitting up front with him while everyone else was sleeping. Windows open, wind blowing, and we were somewhere in the Southwest. His eyes were on the road but his attention was on me, talking through things big and small: God, ancient philosophies and civilizations, plants, and stuff I was interested in, like real life princes, princesses and Kenny Rogers.

> *And your arm felt nice wrapped 'round my shoulder*
> *And I had a feeling that I belonged*
> *I had a feeling I could be someone, be someone, be someone*

I thought of his current situation: seventy-three years old, physically ill, thin, weak, and living in a shack with no running water. Electricity is strung in through an orange extension cord from my sister's house. He intended it to be modeled into a tiny home, but he isn't strong enough to do to the work. We consider hiring the work done for him, but he is stubborn and difficult. It just isn't working out; even the I Ching says so.

> *He says his body's too old for working*
> *I say his body's too young to look like his*
> *My mama went off and left him*
> *She wanted more from life than he could give*
> *I said somebody's got to take care of him*

And I just broke down. I cried because his life is ending, and I'm not sure what his life is about. Maybe the best times were in the blue Ford Econoline van. It's all well and good to be unconventional and rebellious when you're young and strong, but not so pretty when you can't take care of your basic needs.

I'm going to look at some low-income, senior independent living facilities. It'll be grim; that's what $500 per month pays for. But there will be indoor plumbing, running water and a full kitchen. Then, my siblings and I are going to sit down with him and go over his options.

> *You got a fast car*
> *But is it fast enough so you can fly away?*
> *You gotta make a decision*
> *Leave tonight or live and die this way*

I made a map overlapping all the low-income facilities, my house, my sister's house and all the Tulsa area community gardens.

MY LIFE WITH PLANTS | 251

> *And I had a feeling that I belonged*
> *I had a feeling I could be someone, be someone, be someone*
> lyrics from "Fast Car" by Tracy Chapman

Dad arrived at my house in his truck all his worldly possessions and dog. I snapped a picture to share with my siblings.

We unloaded his truck. I suggested he sleep on the couch, or we could put his mattress on living room floor. He opted to sleep on the screened patio. In fact, he pretty much moved into the patio, bringing the mattress, his clothes and a few dishes. He put some foodstuffs into our fridge. His dog, Lucky, stayed at his side or waited patiently by the truck.

By day three, neighbors were warily eyeing Lucky jaunting out and about in the front yard sans leash. Additionally, one could make out odds and ends of clothing and bedding strung here and there as if a vagrant had set up camp. Dad said he loved it on the patio and with easy access to a kitchen and bathroom, it was paradise. He mentioned that if he ever had to have a hospice bed, he would like it to be on the patio.

We began house hunting for him. I drove us around and we toured a few modest rentals. He decided on the first one we saw: a three-room house in a shady (lots of trees and lots of houses with barred windows and doors) neighborhood with a large, fenced yard for $495 a month. Perfect as could be in his budget, not too far from my house and fitting for one man and one dog. After signing the lease, dad wasn't sure he wanted to sleep at his new house that night. He mentioned something about staying at our house for a week. We had been planning a short getaway. I said he could stay, but he'd have to sleep inside the house with the dog in the back yard. Dad opted to head to his new place that evening. We helped him move everything over.

I took off work to take dad to his urology appointment. I picked him up after lunch. He was having a few snags at his place; toilet was clogged, but he was excited because the landlord gave him the okay to build a greenhouse. He'd been avoiding Western medicine ever since his first brush with early stage bladder cancer seven years prior. The Budwig Cancer Diet, vibrational frequencies, and supplements suggested by a Native naturopath healer were his treatments of choice. Recently, he agreed to labwork and imaging.

The wait at the urology office was about 45 minutes. The nurse, Priscilla, was friendly enough, smiling at his dad jokes. She catheterized dad for a urine sample before the doctor looked inside his bladder with a scope.

"I see tumors, extensive tumors, very extensive."

He couldn't tell if the tumors were from the bladder or protruding from the prostrate.

Then causally, "We know you have prostate cancer; your PSA is through the roof."

He also said, "Mets to the bone."

I asked more about the bone cancer and the doctor said, "It's extensive, very diffuse," and handed me a copy of the results. I didn't want to be there, but I didn't want to be anywhere else either.

Cleaning out our family cabin on the Oklahoma Canadian River, I bring home a large box of cassette tapes. My siblings and I had fun with them over the years: recording theatrical demonstrations, made-up songs and play. I was excited about what I might find. I still had a trusty handheld cassette device to listen to the tapes.

There was something about a few of them stuck with me for a few days. Tapes from an old dual-cassette answering machine. One cassette held the outgoing message, my mother's voice, clear and concise.

"We are not able to come to the phone right now. Please leave a message."

The other tape held the messages, a small time capsule of sorts. Lots for my then pre-teen brother. His friends calling with a monotone, "Is Graham there?" They were all indistinguishable from one another. One bright, cheery message from my Aunt Jan. Several from dad. He was calling after work from his job in Tulsa, a 70-minute commute home.

"Hello? Is anyone home? Anyone there?" and another, "I'm just leaving now, going to be home late. Not after dark or anything."

It was a hard time for our family. My brother was coping with social complexities of middle school and at least one bully. My baby sister had a first-grade classmate die from spinal meningitis, and she became seriously afraid of illness and disease. The Ebola crisis unfolding on the nightly news didn't help. My mother was working 12-hour night shifts, and it was all she could do to maintain. I hear the strain in my father's voice.

"Anyone there?"

It sounds thin, hollow, sad.

Dad learns I'm taking a memoir writing workshop. He has been writing the story of his life and suggests I submit his work to my group for feedback. I sigh and explain all of the problems with this. He is undeterred, just shrugs lightly and says he trusts my judgment. He has written chapters about each of his four children. A few days later, he brings over the chapter named Arena and leaves it with me. I read it before bed that night. Below is the passage that stuck with me:

"I took a job at Morgan's flowers in Elgin, Illinois, on the beautiful Fox River. Shortly, after I started there, Marsha asked me to leave. I was surprised; it was just that I thought I had been an upright husband and father since I left the enchantment of New Mexico. During this time, Arena was my most important friend. She spent every other weekend with me. I was, as they say, not in a good place, but I was still a good father. Arena and I walked to church Sunday mornings. I don't remember meeting anyone there, but I know everyone was sympathetic. From the beginning of her ability to speak, Arena had been bright and charming. It was apparent that she was of very small stature, but we didn't think she would end up under five feet tall. Never mind, she had a big voice, once breaking a quiet moment with the extended family seated for an Easter dinner exclaiming, 'Jesus Christ! There's a leaf in my salad!' My relationship with my tiny daughter undoubtably convinced Marsha to try again and made possible the loving family I have today. I could say I owe her virtually everything I hold dear."

In return for my suffering, cancer has given me a whole and wonderful new world to inhabit. It is like a bubble of love, almost a complete harmony. I experience no negative energy. Anger seems no longer to exist in my world. I am closer to my children and my family, even Marsha, than ever before. Though my pension is small, my family provides me with total security including grass-cutting and biweekly maid service. All my relationships have deepened, and many new sincere friendships have been appearing. My body is relaxed when I am moving, and at rest, it becomes very relaxed and still. This gives me more time than I can use to also still my mind and summon the emotion. As my personality gives up authority, my essence appears. I find real attention and can ponder and prepare for the unknowable.

Sand Mueller, lives in Tulsa, Oklahoma, near his children, still working with plants.

www.ingramcontent.com/pod-product-compliance
Lightning Source LLC
Chambersburg PA
CBHW070731020526
44118CB00035B/1182